WARRIOR NOTES HOMESCHOOLING

Math 2
2nd Grade
Units 9-17

Warrior Notes Homeschooling

2nd grade Curriculum: Math book 2

A concise and balanced curriculum that will help build a solid foundation in math with God's love at the forefront.

1 John 4:19 TPT
"Our love for others is our grateful response to the love God first demonstrated to us."

1 Corinthians 13:2 TPT
"And if I were to have the gift of prophecy with a profound understanding of God's hidden secrets, and if I possessed unending supernatural knowledge, and if I had the greatest gift of faith that could move mountains, but have never learned to love, then I am nothing."

Table of Contents

Introduction

Scope and Sequence

Unit 9
Weeks 19 and 20 will focus on division

Week 19: Basic concepts of division, symbols, terms

Week 20: Dividing by 1,2,3,5,10, and math stories

Unit 10
Weeks 21 and 22 will focus on fractions

Week 21: Reviewing and comparing fractions and mixed numbers

Week 22: Fraction math stories

Unit 11
Weeks 23 and 24 will focus on decimals

Week 23: Connecting money and fractions, whole and parts

Week 24: Lining up decimals before math operations, using 0 as a place holder.

Unit 12
Weeks 25-26 will focus on money

Week 25: Adding money using math stories.

Week 26: Subtracting money using math stories.

Unit 13
Weeks 27 and 28 will focus on time

Week 27: Hour and minute hand, a.m. and p.m. time vocabulary review

Week 28: Days and weeks in a year, time lapse

Unit 14
Weeks 29 and 30 will focus on measuring

Week 29: Measuring °C and °F, boiling and freezing points, standard temperatures

Week 30: Measuring length, standard and metric

Unit 15
Weeks 31 and 32 will focus on more measuring

Week 31: Measuring weight, standard and metric

Week 32: Measuring capacity, standard and metric

Unit 16
Weeks 33 and 34 will focus on graphs and grids

Week 33: Line graphs, bar graphs, pie charts, and pictographs

Week 34: Reading a grid and map

Unit 17
Weeks 35 and 36 will focus on basic geometry

Week 35: Recognizing and drawing: triangles, quadrilaterals, pentagons, and hexagons, line of symmetry

Week 36: Recognizing and drawing: spheres, cubes, pyramids, identifying the number of vertices in plane and solid shapes

Vocabulary cards

Answer keys

Resources

Unit 9
Week 19-20

Division

Unit 9 Instructions

Welcome to Warrior Notes Grade 2 Math Book 2! We are excited to partner with you in teaching your children at home, and helping you excel in this honorable endeavor while keeping God at the center of all learning.

Unit 9 introduces a new math operation - division. This can be explained as dividing a number into equal groups. Students will practice doing this by drawing items into groups so they can see how it works. Once they fully understand what division is, then they can began memorizing division facts and the new vocabulary words introduced in this unit.

As with all grammar level math lessons, the best way for students to learn is with hands-on activities. You can use building blocks, puzzle pieces, dice, stuffed animals, and other toys to demonstrate most of the lessons in this unit. You can ask them to help you evenly divide a group of items in half, or in thirds.

Continue asking number-related questions throughout the day to your student to help them see that math and counting is everywhere. You can make flashcards together to practice their division facts and take turns quizzing each other. Put an amount of food items out at lunch or dinner, and ask them how they can be divided up evenly between everyone at the table.

Your love of learning math will be contagious. Pray that God will be present in all of your homeschooling lessons and He will equip you to be the best teacher for your child.

Week 19

Introduction to Division

Division is when we divide things into groups.

12 buckets divided into
3 equal groups =
4 buckets in
each group.

$12 \div 3 = 4$

Draw lines to divide the trees into equal groups.
Color each group a different color.

6 trees divided into 3 groups.

8 trees divided into 4 groups.

Dividing

Draw a line to connect the matching group of pictures
with their equation.

$10 \div 2 = 5$

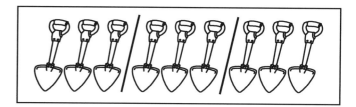

$7 \div 7 = 1$

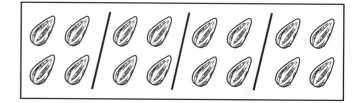

$9 \div 3 = 3$

$16 \div 4 = 4$

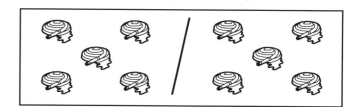

$8 \div 4 = 2$

$6 \div 3 = 2$

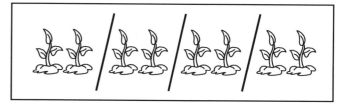

God loves _____ for eternity.

Dividing into Groups

Let's divide this bunch of flowers into 3 equal groups.

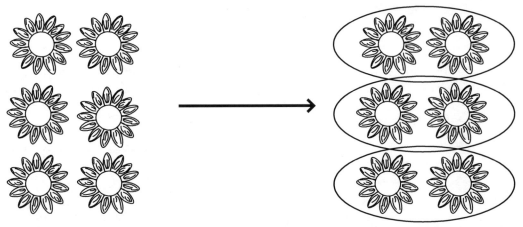

There are 2 in each group.

Divide each set of garden items into equal groups.
How many are in each group?

Draw it: Divide 16 shovels into 2 equal groups.

_____ in each group

Draw it: Divide 10 pots into 5 equal groups.

_____ in each group

Dividing in the Garden

Draw pictures of the total items divided into equal groups.
How many items are in each group?

Divide 12 bags of dirt into 4 equal groups.

12 ÷ 4 = _____

Divide 20 seeds into 2 equal groups.

20 ÷ 2 = _____

Divide 24 buckets into 6 equal groups.

24 ÷ 6 = _____

Divide 30 sprinklers into 5 equal groups.

30 ÷ 5 = _____

Divide 15 hoses into 3 equal groups.

15 ÷ 3 = _____

The Lord is faithful to _____.

Divide and Multiply

For each picture, fill in the blanks with
how many items are in each group.

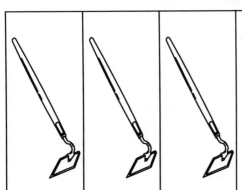

___4___ x _____ = ___4___

___4___ ÷ ___4___ = _____

___4___ x _____ = ___12___

___12___ ÷ ___4___ = _____

___2___ x _____ = ___8___

___8___ ÷ ___2___ = _____

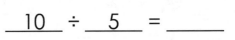

___5___ x _____ = ___10___

___10___ ÷ ___5___ = _____

Division Mix-Up

The numbers for the division problems were written out of order.
Rewrite them in the correct order to show:
total items ÷ number of groups = items in each group.
Hint: the biggest number is the total number and comes first.

6	24	4

_____ ÷ _____ = _____

4	8	2

_____ ÷ _____ = _____

2	18	9

_____ ÷ _____ = _____

5	15	3

_____ ÷ _____ = _____

4	16	4

_____ ÷ _____ = _____

7	2	14

_____ ÷ _____ = _____

3	7	21

_____ ÷ _____ = _____

5	30	6

_____ ÷ _____ = _____

3	4	12

_____ ÷ _____ = _____

1	7	7

_____ ÷ _____ = _____

10	3	30

_____ ÷ _____ = _____

5	20	4

_____ ÷ _____ = _____

God helps _____ through every trouble.

Division Vocabulary

Here are two different ways to write a division equation.
There are three parts to every division equation.

The first number in a division equation is the total number of items that are going to be divided. It is called the **dividend**.

$$24 \div 3 = \underline{8}$$

$$\frac{24}{3} = \underline{8}$$

The second number is the divisor.

The third number or the answer is called a quotient.

Fill in the blanks to show both ways to write the same division equation. Circle the quotients.

$$12 \div 3 = 4$$

$$\frac{12}{\boxed{}} = 4$$

$$10 \div 2 = \boxed{}$$

$$\frac{10}{2} = 5$$

$$\boxed{} \div 5 = 3$$

$$\frac{15}{5} = \boxed{}$$

$$8 \div \boxed{} = 4$$

$$\frac{\boxed{}}{2} = 4$$

Repeated Subtraction

To divide, we can subtract the same amount until there is 0 left.

$12 \div 3$ $\dfrac{12}{-3} = 9$ $\dfrac{9}{-3} = 6$ $\dfrac{6}{-3} = 3$ $\dfrac{3}{-3} = 0$

We subtracted four times. So, the answer to $12 \div 3$ is 4.

Subtract to find the quotients.

$15 \div 5$ $\dfrac{15}{-5}$ = ____ $\dfrac{\boxed{}}{-5}$ = ____ $\dfrac{\boxed{}}{-5}$ = ____

How many times did you subtract? ____ $\qquad$ $15 \div 5$ = ____

$20 \div 4$ $\dfrac{20}{-4}$ = ____ $\dfrac{\boxed{}}{-4}$ = ____ $\dfrac{\boxed{}}{-4}$ = ____ $\dfrac{\boxed{}}{-4}$ = ____ $\dfrac{\boxed{}}{-4}$ = ____

How many times did you subtract? ____ $\qquad$ $20 \div 4$ = ____

$16 \div 8$ $\dfrac{16}{-8}$ = ____ $\dfrac{\boxed{}}{-8}$ = ____

How many times did you subtract? ____ $\qquad$ $16 \div 8$ = ____

$24 \div 6$ $\dfrac{24}{-6}$ = ____ $\dfrac{}{-6}$ = ____ $\dfrac{}{-6}$ = ____ $\dfrac{}{-6}$ = ____

How many times did you subtract? ____ $\qquad$ $24 \div 6$ = ____

The Lord has planned out a future for _____.

Division by 2

Dividing by 2 is dividing a group of items in half.

$$4 \div 2 = 2$$

For each equation below, draw 2 circles to show the two halves.
Draw the same amount in each. The first one is done for you.

$10 \div 2 = 5$

$16 \div 2 = 8$

$20 \div 2 = 10$

$8 \div 2 = 4$ $12 \div 2 = 6$ $14 \div 2 = 7$ $4 \div 2 = 2$

$18 \div 2 = 9$ $2 \div 2 = 1$ $22 \div 2 = 11$ $6 \div 2 = 3$

Color by Number - Division

Color each section by their quotients.
Now draw yourself in the picture!

3 white   4 blue   5 green   2 brown
          1 gray   6 orange   7 aqua

$12 \div 2$

$6 \div 2$

$8 \div 2$

$10 \div 2$

$4 \div 2$

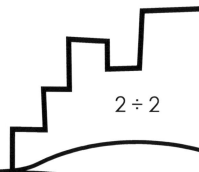
$2 \div 2$

$10 \div 2$

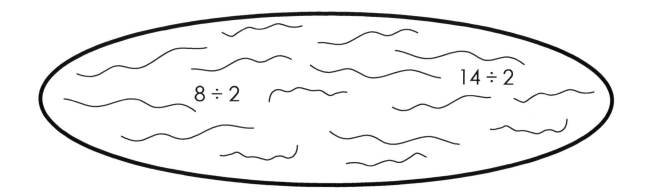
$8 \div 2$ $14 \div 2$

Week 20

Division Recap

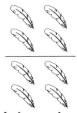

Division is splitting things into equal sized groups.

Quotients are the answers to division problems.

Dividends are the total numbers that are divided into groups.

We can divide by using repeated subtraction.

Dividing by 2 is dividing a group in half.

Find some things in your room or your house that you can divide into groups. What did you find? How many are there total?

_____.

Divide these things into equal groups. How many groups do you have?

_____.

Count how many items are in each group. They should all be the same amount. How many items do you have in each group?

_____.

Write a division equation showing all your work.

_____ ÷ _____ = _____

    # items total   # of groups   # of items in
                                     each group

Division Recap

Draw a line to connect each division fact
with its matching example.

Division is
repeated
subtraction.

$$24 \div 6 = 4$$

Division is
splitting things
into equal sized
groups.

$$35 \div 7 = 5$$

Quotients are
the answers
to division
problems.

$$30 \div 6$$

30	24	18	12	6
- 6	- 6	- 6	- 6	- 6
24	18	12	6	0

Dividends are
the total numbers
that are divided
into groups.

$$18 \div 2 = 9$$

Dividing by 2
is dividing a
group in half.

16 can be split into
4 groups of 4.

Jesus gave His life so _____ could have eternal life.

Division Math Stories

In a math story, we can look for special words
that tell us to **divide** into **groups**.

The chore list was **split** into **equal** jobs for everyone.

The students will **share** the pencils **evenly**.
Find out how many pencils **each** will get.

Read each story and underline the words that show division.
Copy the division equation.

When the girls cleaned their room, they divided the 16 shirts into
4 groups. How many are in each group?

$16 \div 4 = 4$ _____

Wendi and Gary put the 20 books into 5 piles. How many
books are in each pile?

$20 \div 5 = 4$ _____

The 12 moving boxes were filled and put on the truck in 3 rows.
How many boxes are in each row?

$12 \div 3 = 4$ _____

Mom made 8 different bowls of modeling clay. The 2 brothers
shared it all equally. How many did each one get?

$8 \div 2 = 4$ _____

More Division Stories

Draw a picture that shows how each math story is divided.
The first one is done as an example.

The kids put 9 hexagons into 3 equal groups.

The 5 second-graders share 20 squares equally.

Each of the 8 tables has the same amount of triangles. There are 16 total.

There are 18 circles to be divided into 3 groups evenly.

The 12 rectangles are split into 4 equal groups.

The class divided 15 stars between 5 students.

10 ovals need to be shared evenly by 2 friends.

God has planned a great future for _____.

Dividing into 3 Groups

To divide things into equal groups, we can draw an empty space for each group and add 1 to each group at a time, until they each have the exact same amount.

The 3 children earned $15 for weeding the garden.
How many dollars do they each get?

After adding 1 dollar at a time to each group,
we finish with 5 dollars for each child. $15 ÷ 3 = $5

The 9 puppies have 3 dog beds to sleep on. How many puppies will sleep on each bed? Draw them in the beds below.
Make sure each bed has the same amount.

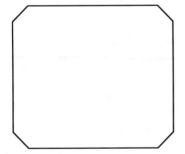 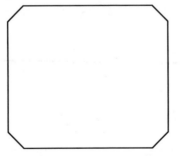 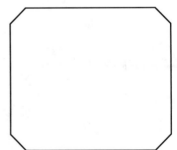

How many dogs are in each bed? _____

We write this as: 9 ÷ 3 = _____

Dividing by 3

Divide up the totals by drawing sketches or tally marks into each empty space as described in the math stories.

Three friends want to share a package of 30 fruit snacks.
Sketch them on their plates.
Make sure each plate has the same amount.

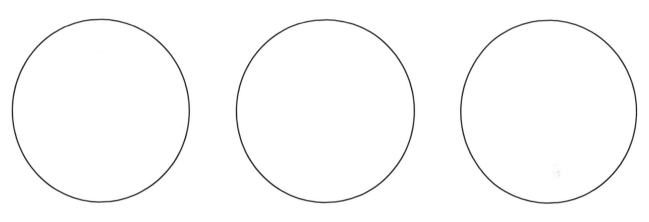

How many fruit snacks does each friend have? _____

We write this as: 30 ÷ 3 = _____

The Sunday School class is ready to color a picture of Acts 2 and the tongues of fire. There are 60 crayons in the box and 3 tables of children. Sketch the crayons onto the tables. Make sure each table has the same amount.

How many crayons does each table have? _____

We write this as: 60 ÷ 3 = _____

Division Word Stories

If Jesus' disciples go fishing with 6 worms for 2 fishing poles, they can divide them equally and use 3 worms per pole. Let's write this same equation in two different ways.

$$6 \div 2 = 3 \qquad\qquad \frac{6}{2} = 3$$

Read each story and show the division equation in two ways.

The book of Daniel in the Old Testament has 12 chapters. If Maddie wants to read the whole book in 2 days, how many chapters will she read each day?

_____

If Daniel had 10 dreams from God and he had 2 each night, how many nights did he dream?

_____

If there were 20 birds on the ark, half of them were male. How many male birds were there?

_____

If David saved 14 of his sheep from wolves during 2 days of shepherding. How many did he save each day?

_____

Division Review

The total number to be divided up is called the **dividend**.

The number it is divided by is called the **divisor**.

The answer to a division problem is called the **quotient**.

For each division equation below, circle the dividend, make a square around the divisor, and underline the quotient.

$24 \div 3 = 8$ $\qquad \dfrac{24}{3} = 8$

$10 \div 5 = 2$ $\qquad \dfrac{10}{5} = 2$

$21 \div 3 = 7$ $\qquad \dfrac{21}{3} = 7$

$25 \div 5 = 5$ $\qquad \dfrac{25}{5} = 5$

$15 \div 3 = 5$ $\qquad \dfrac{15}{3} = 5$

$5 \div 5 = 1$ $\qquad \dfrac{5}{5} = 1$

$9 \div 3 = 3$ $\qquad \dfrac{9}{3} = 3$

$15 \div 5 = 3$ $\qquad \dfrac{15}{5} = 3$

The Lord chose _____ to be His child.

Multiplying and Dividing by 1

If we have 5 dice and put them into 1 group, that group has all 5 dice. 5 total dice divided into 1 group equals 5 dice.

$$5 \div 1 = 5$$

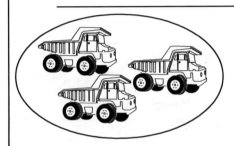

If we have 3 play trucks and put them into 1 group, there is 1 group of 3.

$$3 \div 1 = 3$$

$4 \times 1 = 4$ $4 \div 1 = 4$ $3 \times 1 = 3$ $3 \div 1 = 3$

The **identity property** is a law for multiplication, and it is also true for division. Practice multiplying and dividing by 1 in the examples below.

$\frac{5}{1} =$	$\frac{6}{1} =$	$2 \div 1 =$	$7 \times 1 =$
$12 \div 1 =$	$10 \times 1 =$	$\frac{8}{1} =$	$3 \times 1 =$
$4 \times 1 =$	$9 \div 1 =$	$14 \div 1 =$	$\frac{11}{1} =$

Dividing into Groups of 3, 4, 5

 Use tally marks to sketch the group sizes given. Then, count how many groups you have total.

Selah baked 30 cookies for her neighbors. She put 5 cookies on each plate to give away. Draw the plates below with 5 cookies on each one.

How many plates did she give away? _____ 30 ÷ 5 = _____

Seth lines up 4 of his toy cars on each bookshelf. He has 20 cars in all. Draw the shelves below with 4 cars on each one.

How many shelves will he use for his cars? _____ 20 ÷ 4 = _____

Cambri finished her tennis lesson and put 3 tennis balls in each can. She had 21 tennis balls total. Draw the cans below with 3 balls in each.

How many cans will she use for the tennis balls? ____ 21 ÷ 3 = ____

Unit 10
Week 21-22

Fractions

Unit 10 Instructions

Unit 10 teaches fractions as part of a whole. New vocabulary words are introduced and should be learned after the concept of fractions is understood. You can show them how the fraction looks just like the division problems from the last unit. Show them that a fraction is a number of equal-sized groups, just like division is putting things into equal sized groups.

A simple way to learn fractions is by slicing 2 pizzas with different sized slices. One can be sliced into 4 pieces, and one can be sliced into 8 pieces. You can show them a slice from the first pizza is 1/4 but a slice from the second pizza is 1/8. This can be done on paper, or with real food. Have them draw a picture of a pizza and color it in. Then they can cut it into their choice of equal-sized slices and label each one with the correct fraction.

Continue asking number-related questions throughout the day to your student to help them see that math and counting is everywhere. You can make flashcards together to practice their new vocabulary and take turns quizzing each other. You can divide up a group of cars or other toys into 3 equal sized groups and then ask them to show you 1/3 of the toys. Then you can put the toys back together and divide them into 4 equal-sized groups. Ask them to show you 1/4 of the toys and then 1/2 of the toys. The main idea to take away is that a fraction is always a part of a whole. A 1/2 of a dollar is only part of a dollar, and not a whole dollar.

Your love of learning math will be contagious. Pray that God will give you creativity in all of your homeschooling lessons and He will equip you to be the best teacher for your child.

Week 21

Fraction Basics

Fractions are numbers that can show part of an equal amount.

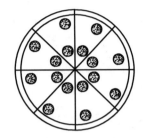  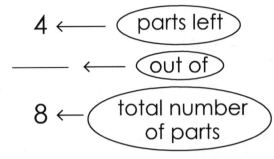

4 ← parts left

___ ← out of

8 ← total number of parts

This is 1 whole pizza

1

How much pizza is left?

$\frac{4}{8}$

We have 4 parts left out of 8 equal parts total.

All of the pizzas started with 8 equal slices.
Finish each fraction by writing in the amount of slices left.

 $\frac{\quad}{8}$

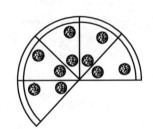 $\frac{\quad}{8}$

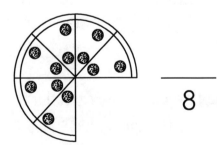 $\frac{\quad}{8}$

 $\frac{\quad}{8}$

 $\frac{\quad}{8}$

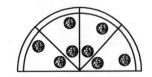 $\frac{\quad}{8}$

Fraction Basics

Circle the correct fraction that matches
the shaded part of each picture.

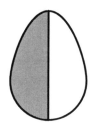

$$\frac{1}{3}$$ $$\frac{1}{4}$$ $$\frac{1}{2}$$

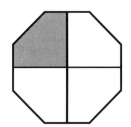

$$\frac{1}{3}$$ $$\frac{1}{4}$$ $$\frac{1}{2}$$

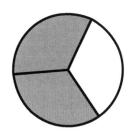

$$\frac{1}{3}$$ $$\frac{1}{4}$$ $$\frac{2}{3}$$

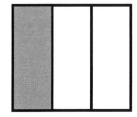

$$\frac{1}{3}$$ $$\frac{1}{4}$$ $$\frac{1}{2}$$

_____ is a part of God's kingdom.

Fractions as Part of a Set

The top number in a fraction is the **numerator**.

$$\frac{1}{3}$$

The bottom number in a fraction is the **denominator**.

One third of the equal stars are shaded.

Color in one third (1/3) of the pictures in each set.

Color in one fourth (1/4) of the pictures in each set.

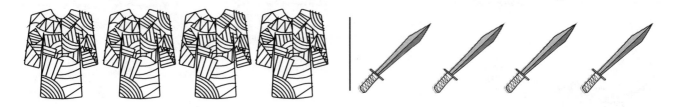

Color in one half (1/2) of the pictures in each set.

Color in one tenth (1/10) of the pictures in each set.

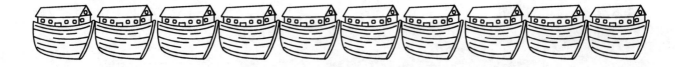

Fractions as Part of a Whole

Color in one fourth (1/4) of each shape.

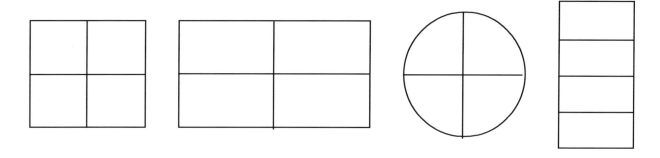

Color in one third (1/3) of each shape.

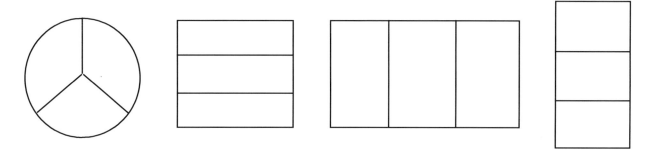

Color in one half (1/2) of each shape.

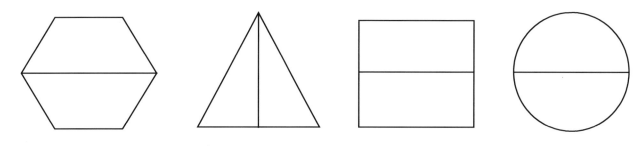

Color in one tenth (1/10) of each shape.

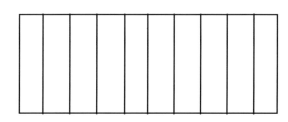
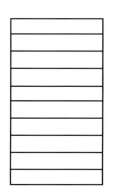

Jesus sticks closer than a brother to _____.

Fractions that Equal 1

This square has all 4 parts shaded.

$$\frac{4}{4} = 1$$

There are 4 parts shaded out of 4 parts total. This fraction is 4/4.

The whole square is shaded. 1 whole square is shaded. When the numerator and denominator match, the fraction = 1.

Write in the fractions to show each amount.

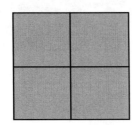

$$\frac{\square}{\square} = 1$$

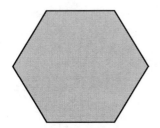

$$1 = \frac{\square}{\square}$$

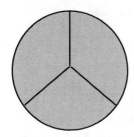

$$\frac{\square}{\square} = 1$$

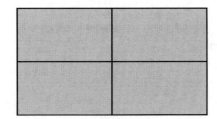

$$1 = \frac{\square}{\square}$$

Fractions that Equal 1

Circle the correct fraction that matches each amount.

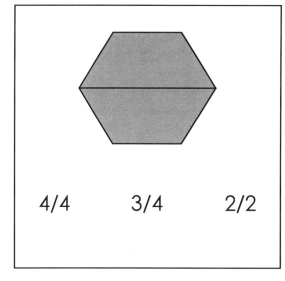

4/4   3/4   2/2

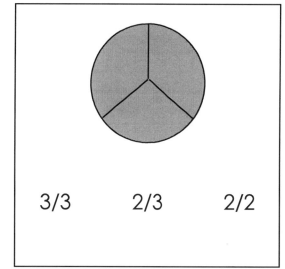

3/3   2/3   2/2

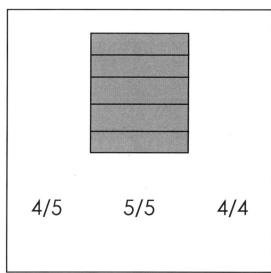

4/5   5/5   4/4

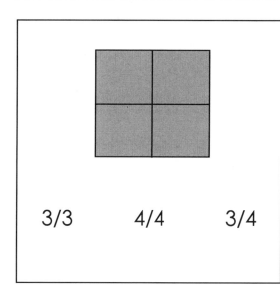

3/3   4/4   3/4

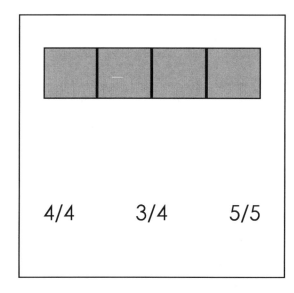

4/4   3/4   5/5

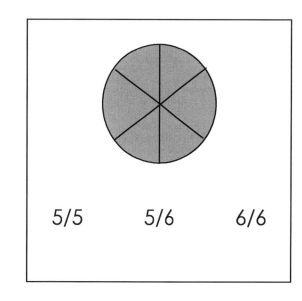

5/5   5/6   6/6

_____ has an amazing future with God.

Mixed Numbers

Sometimes we have whole items AND parts of items. We show these amounts by using mixed numbers.

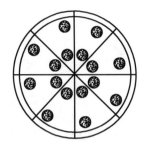 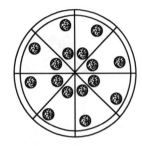 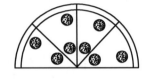 $2\dfrac{1}{2}$

We have 2 whole pizzas, AND we also have one-half (1/2) of a pizza. We write the number of whole pizzas first; then we write the fractional parts. There are two and a half pizzas left.

Color in the pizzas to show each whole and fraction amount.

$1\dfrac{1}{8}$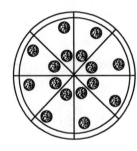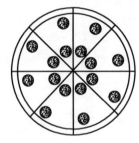

$2\dfrac{6}{8}$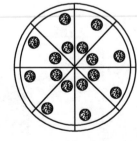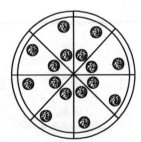

$\dfrac{1}{2}$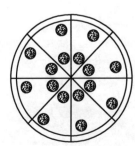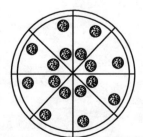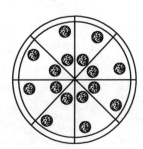

Comparing Fraction Amounts

Use the pictures to decide which amount is greater.
Circle the fraction that is greater.

$$\frac{4}{6} \qquad \frac{1}{3}$$

$$\frac{3}{4} \qquad \frac{1}{2}$$

$$\frac{3}{5} \qquad \frac{2}{4}$$

$$\frac{1}{2} \qquad \frac{2}{3}$$

The Lord loves to hear from _____ every day.

Fractions

Fill in the blanks below each set of pictures.

There are _____ total faces.

_____ of them are shaded.

$\dfrac{\square}{\square}$ of them are shaded.

There are _____ total wrenches.

_____ of them are shaded.

$\dfrac{\square}{\square}$ of them are shaded.

There are _____ total worms.

_____ of them are circled.

$\dfrac{\square}{\square}$ of them are circled.

There are _____ total burgers.

_____ of them are circled.

$\dfrac{\square}{\square}$ of them are circled.

There are _____ total ladders

_____ of them are circled.

$\dfrac{\square}{\square}$ of them are circled.

There are _____ total drums.

_____ of them are circled.

$\dfrac{\square}{\square}$ of them are circled.

Fraction Review

Draw a line to match up each fraction fact with its example.

The numerator is the
top number of a fraction.

$$3 \frac{1}{4}$$

A fraction with
matching numbers
is equal to 1.

$$\frac{⑤}{6}$$

A fraction can be
written with a straight
line or a slanted one.

$$\frac{3}{3} = 1$$

A mixed number
shows some whole amounts
and some fraction amounts.

$$\frac{2}{③}$$

The denominator is
the bottom number
of a fraction.

$$3/4 = \frac{3}{4}$$

Week 22

Half, Thirds, and Fourths

The words HALF, THIRD, and FOURTH represent fractions when we are talking about parts of a whole or parts of a group.

Half of the lemons are rotten.
Half is 1 out of 2 equal groups.

$$\frac{1}{2}$$

$$\frac{1}{3}$$

A third of the oranges are ripe.
A third is 1 out of 3 equal groups.

A fourth of the grapes are green.
A fourth is 1 out of 4 equal groups.
A quarter is 1 out of 4 equal groups.

Draw a line to connect the fraction numbers
with matching fraction words.

one quarter   3/4

two thirds   1/4

three fourths   2/2

two halves   2/3

Fruit Fractions

Write a fraction to describe each sentence.

A third of the fruit stand customers
have children with them.

Half of the fruit stand
is painted red.

They are open for a fourth
of the day.

Two thirds of the customers
have bought fruit here before.

Three fourths of the fruit they
sold was eaten in one day.

Half of the shoppers
today are women.

Two quarters of the workers ate
a piece of fruit while at work.

God always helps _____ with every need.

Fraction Stories

In a math story, there are certain words
that show us to use fractions.

2 **out of** the 5 apples $\dfrac{2}{5}$
were red.

3 **out of** the 4 children $\dfrac{3}{4}$
ate apples.

Write a fraction to show each amount described.

Only 1 out of the 12 disciples
betrayed Jesus.

Only 2 out of the 3 women
drew water from the well.

Out of 8 people,
5 asked for prayer.

There were 8 new followers of Jesus
out of the 11 who heard His invitation.

Only 1 out of the 10 healed
men thanked Jesus.

Fruit Stand Fractions

The fruit stand has lots of different fruits for sale. Fill in the blanks with fractions to show all the amounts.

The grapes are green. Only 7 out of the 8 bunches are ripe. The workers put away the sour bunch. What fraction of the bunches are ripe?

There are 15 apples in the basket for sale, but 4 of them are eaten by a worm. They cannot be sold. What fraction of the apples are left for sale?

Out of the 5 lemons, 1 is rotten. What fraction of the lemons are ripe and on sale today?

There are 12 oranges on the table, but 2 of them are old and not fresh anymore. What fraction of the oranges are fresh?

How many pieces of fresh and ripe fruits are left for sale? _____

Write the fraction amount to show how many of the total fruits for sale are apples.

God is faithful to _____ .

Fractional Words

Color in the correct number of each item.

Four ninths of the fruits of the Spirit end with the suffix **NESS**.

Three fourths of the lessons are about the kingdom of God.

The pastor asked five eighths of the leaders to pray.

Two sevenths of the books are about the fruits of the Spirit.

One fifth of the Bibles were left at the front of the church.

More Fractions

Color in the correct number of parts of a whole to match the fractional amount in each sentence.

Four of the six seats in the front row are filled.

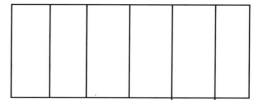

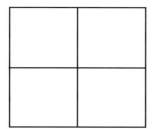

Half of the communion table is covered with bread and juice.

People are praying at three of the four sections of the altar.

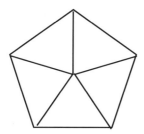

Two out of the five worship songs are new.

One out of the six doors of the church is locked.

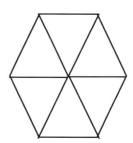

Three out of the eight choir singers are men.

_____ worships the one true and living God.

Matching Fractions

Circle the fraction that matches each picture.

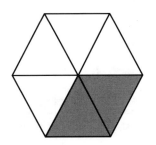

$$\frac{1}{6} \qquad \frac{2}{6} \qquad \frac{6}{2}$$

$$\frac{5}{2} \qquad \frac{3}{5} \qquad \frac{2}{5}$$

$$\frac{2}{3} \qquad \frac{1}{3} \qquad \frac{3}{3}$$

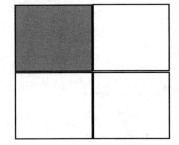

$$\frac{1}{4} \qquad \frac{3}{4} \qquad \frac{4}{4}$$

$$\frac{4}{2} \qquad \frac{2}{4} \qquad \frac{3}{4}$$

Writing Fractions

Write a number fraction for each sentence.
Use the word bank below to fill in the fraction word blank.

Two out of the nine fruits of the Spirit start with P.

$$\frac{\boxed{}}{\boxed{}}$$

_____

One out of four disciples spoke to the crowd.

$$\frac{\boxed{}}{\boxed{}}$$

_____

Eleven out of the fifteen verses were read out loud.

$$\frac{\boxed{}}{\boxed{}}$$

_____

The people followed Jesus for four out of seven days.

$$\frac{\boxed{}}{\boxed{}}$$

_____

Two quarters of the crowd were healed.

$$\frac{\boxed{}}{\boxed{}}$$

_____

Five out of the six kids hugged Jesus.

$$\frac{\boxed{}}{\boxed{}}$$

_____

Word Bank

two-fourths   four-sevenths   two-ninths

five-sixths   eleven-fifteenths   one-fourth

_____ lives by the power of the Holy Spirit.

Fraction Review

Draw a line to connect each fraction to its match.

2/5   One

1/6   Three Quarters

4/4   Two Thirds

2/3   Half

3/4   Four

8/10   One Sixth

4/1   Five Eighths

1/2   Two Fifths

5/8   Eight Tenths

Fraction Story Review

Read the story and answer the questions.
Use the blank space for your work.

There are 32 people at the church prayer meeting and 20 are women. There are 12 people sitting in chairs. The rest are standing or kneeling.

What fraction of people at the prayer meeting are men and children?

What fraction of people are not sitting?

One quarter of the women are wearing a hat. How many women are not wearing a hat?

At 9 PM, 24 people leave to go home. What fraction of people are left at prayer?

What would you pray about at a church prayer meeting?

Unit 11
Week 23-24

Decimals

Unit 11 Instructions

Unit 11 introduces decimals. Since students have just learned about how fractions are parts of a whole, this is an easy transition to see that decimals are just another way to write parts of a whole.

The easiest way to explain decimals is by using coins. Gather as many pennies, dimes, and quarters as you can find and have them count how many dimes are in a dollar. Once they have 10, show them how one dime is 1/10 of a dollar. Ten dimes are 10/10 of a dollar, or one whole dollar. You can then practice with pennies and quarters.

When they learn how to write down an amount of money, they can apply what they learned from place values. The first number after the decimal point is the dimes and the second number is the pennies. After they understand what these numbers show, then they can learn the new vocabulary words tenths and hundredths.

You can practice calling dimes and pennies by the words tenths and hundredths interchangeably. After they are comfortable with tenths and hundredths, then you can move into using quarters and calling them fourths and quarters interchangeably.

The main goal of this unit is for your child to understand that any numbers after the decimal point are parts of a dollar, or an amount. The best start is to learn tenths, hundredths, and quarters. Have them practice matching up these words with 1/10, 1/100, and 1/4, and 0.1, 0.01, and 0.25. This will be a great foundation for all future math lessons to come.

Your love of learning math will be contagious. Pray that God will help you make homeschooling fun for your children - this will foster a love for learning in their hearts.

God speaks to _____ through His Word.

Introduction to Decimals

A decimal point is written between whole dollars and parts.

Whole dollars | Parts of a dollar

$ 4 . 2 5

The first number after the decimal shows dimes which are **tenths** of a dollar.

The second number after the decimal shows pennies which are **hundredths** of a dollar.

2 dimes and 5 pennies = 25¢
25¢ is equal to one quarter, or 1/4 of a dollar.

Draw an arrow to divide each dollar amount between whole dollars and parts of a dollar. Use the example above.

$ 5 . 7 1 $ 9 . 1 8

$ 1 . 3 2 $ 2 . 9 9

$ 3 . 4 5 $ 8 . 2 7

$ 6 . 8 3 $ 7 . 5 0

Decimal Place Values

Write each dollar amount into the correct place value spaces.

Dollar Amount	Dollars	Tenths	Hundredths
$ 4 . 3 8	4	3	8
$ 9 . 1 5			
$ 3 . 9 1			
$ 1 . 8 9			
$ 8 . 7 7			
$ 2 . 2 9			
$ 5 . 4 6			
$ 7 . 3 4			

Decimals - Whole and Parts

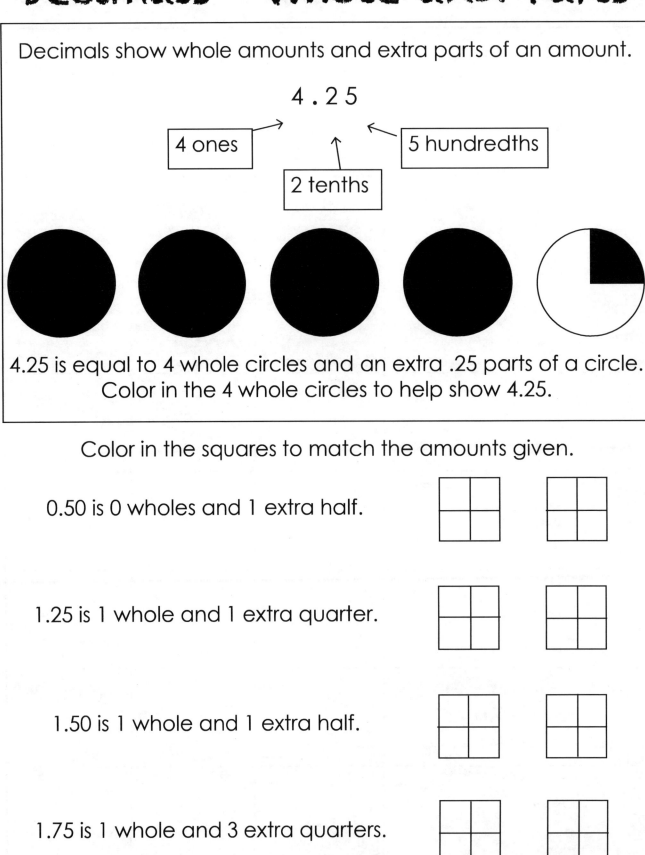

Decimals show whole amounts and extra parts of an amount.

4 . 2 5

4 ones

2 tenths

5 hundredths

4.25 is equal to 4 whole circles and an extra .25 parts of a circle.
Color in the 4 whole circles to help show 4.25.

Color in the squares to match the amounts given.

0.50 is 0 wholes and 1 extra half.

1.25 is 1 whole and 1 extra quarter.

1.50 is 1 whole and 1 extra half.

1.75 is 1 whole and 3 extra quarters.

Decimals - Whole and Parts

Color in the given whole pies and extra parts of a pie.

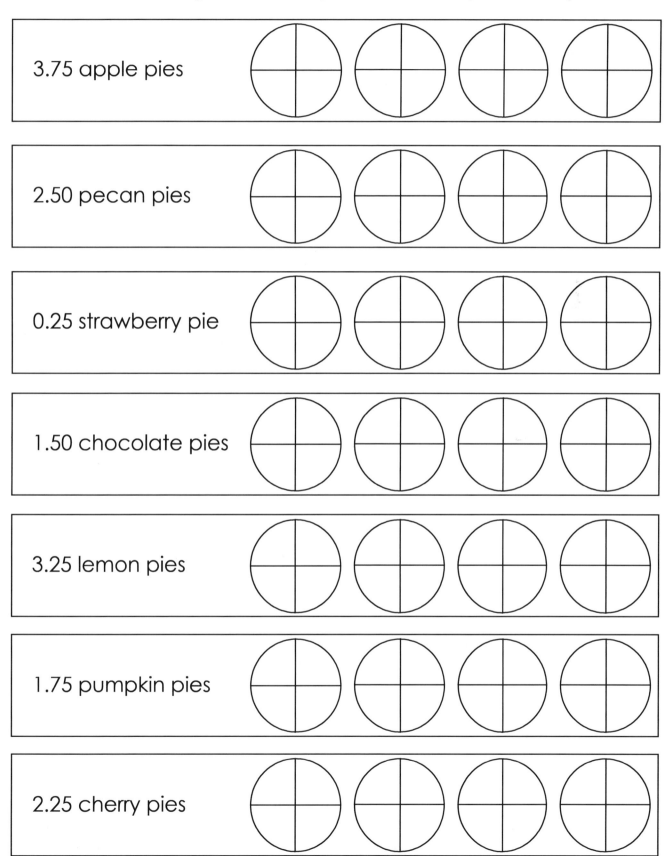

3.75 apple pies

2.50 pecan pies

0.25 strawberry pie

1.50 chocolate pies

3.25 lemon pies

1.75 pumpkin pies

2.25 cherry pies

God created _____ with a future and a hope.

Comparing Decimals

We can compare decimal amounts using >, <, and = symbols.

7.42 > 7.40	5.15 < 5.19	8.38 = 8.38
.42 is greater than .40	.15 is less than .19	.38 equals .38

Write in the correct symbol to compare the decimal amounts.

25.62	☐	25.10

11.55	☐	10.92

67.12	☐	65.14

33.21	☐	33.83

79.42	☐	70.45

84.67	☐	84.67

42.26	☐	62.42

93.52	☐	92.90

52.51	☐	52.51

0.19	☐	1.00

3.81	☐	3.18

17.32	☐	71.32

Number Line Decimals

Write the given amounts in the correct blanks under each number line. Mark a line for each of these decimals.

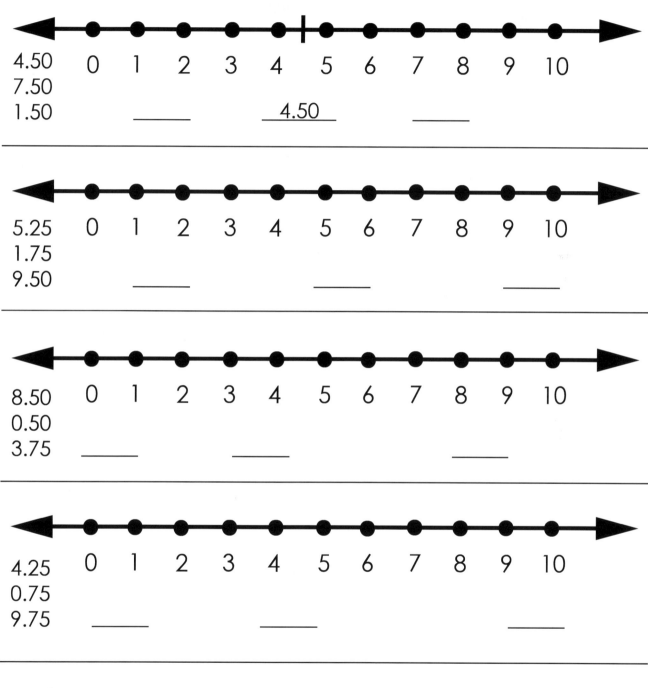

4.50
7.50
1.50

_____   4.50   _____

5.25
1.75
9.50

_____   _____   _____

8.50
0.50
3.75

_____   _____   _____

4.25
0.75
9.75

_____   _____   _____

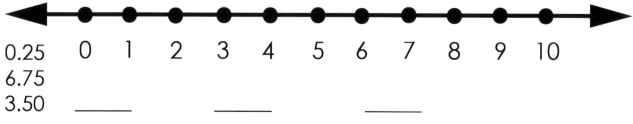

0.25
6.75
3.50

_____   _____   _____

Decimal and Fraction Amounts

We can write the same amount of parts of a whole using both fractions and decimals.

$1\dfrac{3}{4}$  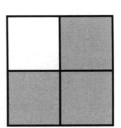 1.75

1 whole and
3 out of 4 equal parts

1 dollar and 75 cents
(3 quarters)

For each picture, write in the fraction and decimal amounts.

Fraction	Decimal
1 1/2	1.5 0

Decimal and Fraction Amounts

Draw a line to connect the decimals with matching fractions.

3.50 $1\dfrac{3}{4}$

0.75 $1\dfrac{1}{4}$

3.25 $3\dfrac{1}{2}$

1.75 $\dfrac{1}{2}$

3.75 $\dfrac{3}{4}$

2.50 $3\dfrac{1}{4}$

0.50 $3\dfrac{3}{4}$

1.25 $2\dfrac{1}{2}$

The Lord blesses _____ with His presence.

Decimals Review

For each number below, there is 1 underlined digit.
Circle whether it is the tens, ones, tenths, or hundredths place.

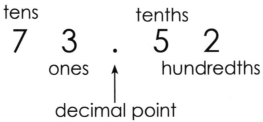

6<u>3</u>.45	tens	ones	tenths	hundredths

<u>1</u>0.29	tens	ones	tenths	hundredths

37.5<u>1</u>	tens	ones	tenths	hundredths

14.<u>5</u>7	tens	ones	tenths	hundredths

28.8<u>9</u>	tens	ones	tenths	hundredths

45.9<u>6</u>	tens	ones	tenths	hundredths

Decimals Review

Rewrite the decimal amounts from each set
in order from least to greatest.

34.23	_31.90_
32.86	_____
39.12	_____
31.90	_____

50.05	_____
51.90	_____
50.75	_____
50.98	_____

14.00	_____
19.60	_____
17.45	_____
14.11	_____

91.36	_____
92.60	_____
90.90	_____
91.63	_____

12.09	_____
21.90	_____
12.00	_____
20.92	_____

65.16	_____
60.56	_____
62.10	_____
65.65	_____

Week 24

_____ reads the Bible every day.

Decimals

Fill in the blanks for each decimal comparison.
There are many correct answers.

5.67 > 5. _____

6.17 < 6. _____

1.90 < 1. _____

3.12 > 3. _____

8.09 = 8. _____

2.24 < 2. _____

4.87 < 4. _____

9.25 > 9. _____

Whole and Part Decimals

Color in the whole and part amounts to match each decimal.

1.75

0.25

2.50

3.25

1.50

God protects _____ as a Heavenly Father.

Decimal Tenths

Every dime is equal to 1 tenth of a dollar.
We can add up money and dimes like this:

$1.00 $1.10 $1.20 $1.30 $1.40 $1.50

$1.60 $1.70 $1.80 $1.90 $2.00

Now, it's your turn! Find as many dimes as you can from your piggy bank or around your house. Each of your dimes is 1 tenth of a dollar. If you have more than 10 dimes, separate them into groups of 10. Now, you can answer the questions below.

How many dimes do you have altogether?   _____

Do you have more than 10?   _____

If you have more than 10, how many groups of 10 do you have in total? These are your dollars. Write this number next to the dollar sign below. If you don't have a full group of 10, write a 0 on the line next to the dollar sign below.

$ _____

How many dimes do you have that aren't in a full group of 10? These are your tenths. Write this number after the decimal point.

$ _____ . _____

How much money do you have in all? $ _____ . _____

Tenths

Color in the given amount of tenths in each block of ten below.
Fill in the blanks and read aloud to show how many tenths.

0.60 = .6

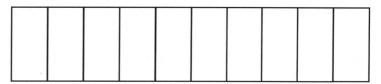

This is _____ tenths.

0.10 = .1

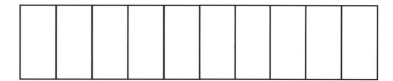

This is _____ tenths.

0.80 = .8

This is _____ tenths.

0.50 = .5

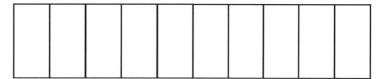

This is _____ tenths.

0.30 = .3

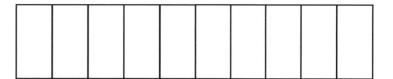

This is _____ tenths.

The Lord is kind to _____.

Decimal Hundredths

Every penny is equal to 1 hundredth of a dollar.

Some amounts have both dimes and pennies.
$4.82

Some amounts only have pennies.
$4.02

Fill in the chart for each amount to show how many whole dollars, how many tenths, and how many hundredths.

	Dollars - Whole	Dimes - Tenths	Pennies - Hundredths
$1.94			
$2.27			
$5.03			
$0.90			

Rounding to Tenths

We can estimate or round hundredths to the nearest tenth.

With 4 or less hundredths, we round down.

↓ 8.34 rounds down to 8.3

With 5 or more hundredths, we round up.

↑ 3.39 rounds up to 3.4

Help the squirrels estimate how many nuts and pieces of nuts they collected this month. Round each amount up or down to the nearest tenth.

5.29 → 5.3

6.89 →

2.13 →

9.74 →

3.91 →

7.55 →

1.48 →

4.62 →

God forgives all the sins of _____.

Adding Decimals

If we need to add 2 amounts, it is very important to line up the decimal points. This way, we are adding
whole numbers to whole numbers, tenths to tenths,
and hundredths to hundredths.

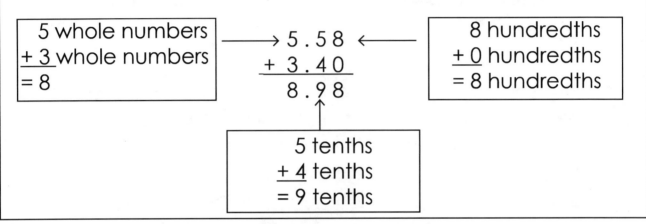

| 5 whole numbers |
| + 3 whole numbers |
| = 8 |

→ 5.58 ←
  + 3.40
   8.98

| 8 hundredths |
| + 0 hundredths |
| = 8 hundredths |

| 5 tenths |
| + 4 tenths |
| = 9 tenths |

Add up the decimal amounts,
starting with the hundredths column and moving to the left.

$$
\begin{array}{r} 15.34 \\ + \ 34.15 \\ \hline \end{array}
\qquad
\begin{array}{r} 22.81 \\ + \ 14.13 \\ \hline \end{array}
\qquad
\begin{array}{r} 19.22 \\ + \ 20.72 \\ \hline \end{array}
$$

$$
\begin{array}{r} 53.43 \\ + \ 23.11 \\ \hline \end{array}
\qquad
\begin{array}{r} 14.16 \\ + \ 52.81 \\ \hline \end{array}
\qquad
\begin{array}{r} 13.65 \\ + \ 60.04 \\ \hline \end{array}
$$

$$
\begin{array}{r} 25.27 \\ + \ 64.30 \\ \hline \end{array}
\qquad
\begin{array}{r} 7.89 \\ + \ 31.00 \\ \hline \end{array}
\qquad
\begin{array}{r} 34.59 \\ + \ \ \ 2.20 \\ \hline \end{array}
$$

$$
\begin{array}{r} 36.53 \\ + \ 41.20 \\ \hline \end{array}
\qquad
\begin{array}{r} 10.73 \\ + \ 25.06 \\ \hline \end{array}
\qquad
\begin{array}{r} 1.94 \\ + \ 50.01 \\ \hline \end{array}
$$

Adding Tenths and Hundredths

Fill in the blanks to add up the hundredths first, the tenths next, and then the whole numbers. Write the final sum in the box.

4.15 + 2.82

____ hundredths + ____ hundredths = ____ hundredths

____ tenths + ____ tenths = ____ tenths

____ whole + ____ whole = ____ whole

1.71 + 3.07

____ hundredths + ____ hundredths = ____ hundredths

____ tenths + ____ tenths = ____ tenths

____ whole + ____ whole = ____ whole

5.69 + 2.1

____ hundredths + ____ hundredths = ____ hundredths

____ tenths + ____ tenths = ____ tenths

____ whole + ____ whole = ____ whole

2.24 + 8.53

____ hundredths + ____ hundredths = ____ hundredths

____ tenths + ____ tenths = ____ tenths

____ whole + ____ whole = ____ whole

_____ will live for eternity with Christ.

Decimal Review

Choose the correct answer for each box.

4.3 + 5.2 =	6.14 ____ 6.4	1 0 2 . <u>7</u> 3
9.05	>	The underlined number =
9.5	<	tens
10.0	=	hundredths
		tenths

2.8 ____ 2.08	1.17 round to tenths	1.05 + 3.1 =
>	1.1	4.6
<	1.2	4.06
=	1.0	4.15

8 1 . 2 <u>6</u>	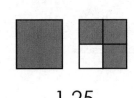	3.83 round to tenths
The underlined number =	1.25	3.8
tens	1.50	3.9
hundredths	1.75	4.0
tenths		

Adding Decimal Review

Circle the two amounts that can be added together to equal the given sum.

| 2.05 | 2.10 | 3.2 | → | 5.25 |

| .5 | 1.02 | 1.2 | → | 1.7 |

| 2.05 | 1.4 | 1.04 | → | 3.09 |

| 1.5 | 1.04 | 1.40 | → | 2.90 |

| 1.25 | 1.50 | 3.5 | → | 4.75 |

| 0.75 | 0.50 | 0.3 | → | 1.05 |

Unit 12
Week 25-26

Money

Unit 12 Instructions

Unit 12 focuses on adding and subtracting with decimals. The most important takeaways from this unit are to line up decimal points before starting any math operations, and to always put a zero in as a place holder for any missing digits. A 4 has the same value as 4.0 which has the same value as 4.00 and so on.

The unit continues to build and practice on the grouping and regrouping processes which were learned in first grade and the first book in second grade. You can show your student that these processes don't change when decimals are introduced. All of the addition and subtraction rules still apply when decimals are included in a number. This unit also reinforces the idea that decimals are only part of a dollar, and anything to the right of a decimal point shows part of a whole.

A new concept that they haven't seen before is regrouping more than once. So they will regroup to make 10, and then regroup from that 10 which will change it to 9. For example: 200 - 7 done as a vertical subtraction problem will require the student to regroup from the hundreds column and add a 1 to the tens column and make 10 tens. Next, they will have to regroup from this tens column and add a 1 to the ones column. Their 10 tens is now 9 tens. Now they can subtract the 7 from the 10 in the ones column. Using a white board and writing large numbers may be a easier way to show your child how this process works. If they grasp the concept, you can move to the next step of subtracting 1000 - 8 and show them how to regroup three times. Take all the time needed and allow them a break in between the two math pages for each day if they need it. Our lessons are set up as two different but complimentary pages per day. Your child might want to do both pages at once, or complete the first page in the morning and the second page as homework later in the day. Ask God to give you wisdom about the best way for your child to learn these important concepts.

_____ reads the Word of God.

Adding Decimals in Money

When adding decimals, always line up the decimal points first.

$$
\begin{array}{r}
\$\ 3\ 1\ 4.7\ 3 \\
+\$\ 2\ 6\ 2.2\ 4 \\
\hline
\$\ 5\ 7\ 6.9\ 7
\end{array}
$$

$$
\begin{array}{r}
{\scriptstyle 1} \\
\$\ 4\ 7\ 2.5\ 2 \\
+\$\ 1\ 2\ 5.3\ 9 \\
\hline
\$\ 5\ 9\ 7.9\ 1
\end{array}
$$

2 + 9 = 11
Move the ten to the tens column.

The decimals are lined up. Now add to find the sums in each piggy bank.

$$
\begin{array}{r}
\$1\ 5\ 3.6\ 5 \\
+\$2\ 2\ 5.1\ 4 \\
\hline
\end{array}
$$

$$
\begin{array}{r}
\$9\ 6\ 2.2\ 8 \\
+\$\ \ \ 2\ 4.1\ 1 \\
\hline
\end{array}
$$

$$
\begin{array}{r}
\$7\ 4.8\ 3 \\
+\$\ \ \ 5.1\ 5 \\
\hline
\end{array}
$$

$$
\begin{array}{r}
\$9\ 1.4\ 2 \\
+\$\ \ \ 7.3\ 7 \\
\hline
\end{array}
$$

$$
\begin{array}{r}
\$3\ 7\ 1.5\ 4 \\
+\$3\ 5\ 3.2\ 3 \\
\hline
\end{array}
$$

$$
\begin{array}{r}
\$4\ 3\ 3.7\ 1 \\
+\$1\ 2\ 4.0\ 5 \\
\hline
\end{array}
$$

$$
\begin{array}{r}
\$1\ 0\ 8.3\ 6 \\
+\$\ \ \ 7\ 1.5\ 6 \\
\hline
\end{array}
$$

$$
\begin{array}{r}
\$2\ 4\ 4.9\ 2 \\
+\$\ \ \ 3\ 2.3\ 3 \\
\hline
\end{array}
$$

Addition with Decimals

Rewrite the amounts and line up the decimals.
Add to find the sums.

$42.50 + $31.32

+ _____

= _____

$3.49 + $90.17

_____

+ _____

= _____

$29.64 + $5.24

_____

+ _____

= _____

$68.03 + $171.86

_____

+ _____

= _____

God is a loving Father to _____.

Picnic Math Stories

Solve the math stories. Line up decimal points before adding.

The picnic basket cost $7.50 and the plates for lunch cost $13.20. How much was spent for the picnic?

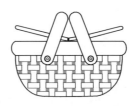

_____
+ _____
= _____

The apples cost $8.75 and the berries cost $3.80. How much was spent on fruit?

_____
+ _____
= _____

The bread cost $5.36 and the jam cost $3.21. What is the total?

_____
+ _____
= _____

The chicken nuggets cost $11.47 and the fries cost $4.59. What is the total?

_____
+ _____
= _____

The cupcakes cost $9.18 and the cobbler cost $2.95. How much was spent on dessert?

_____
+ _____
= _____

Decimal Addition

Solve each math addition story by lining up the decimal points. Then, round each amount to the closest dollar.

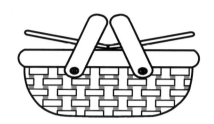

After the picnic, we paid $15.60 for adult paddle boat tickets and $8.25 for child tickets. What was our total?

$15.60
+$ 8.25
$23.85

Rounds to:

_____

The mini golf was $5.90 for the kids and $12.85 for the adults. What was the total mini golf cost?

Rounds to:

+ _____

_____

The parking lot cost $5.75 for the first hour and then $3.50 for the next hour. What was the total spent on parking?

Rounds to:

+ _____

_____

The kids won $8 in the golf contest and then found $4.20 in change around the park. How much money was it in all?

Rounds to:

+ _____

_____

They used $1.97 of bread and $2.44 of crackers to feed the fish in the lake. What was the total?

Rounds to:

+ _____

_____

The Lord reveals Himself to _____.

Decimals in Money Stories

Solve the math stories by adding. Then write
in the number of dollars and coins needed for the total.

The kindergarten students saved $4.25 from their allowances and earned $11.87 in their bake sale fundraiser. How much did they have total?	Whole dollars _____ Quarters _____ Dimes _____ Nickels _____ Pennies _____
The first grade students saved $6.15 from their allowances and earned $10.52 in their dog walking fundraiser. How much did they have total?	Whole dollars _____ Quarters _____ Dimes _____ Nickels _____ Pennies _____
The second grade students saved $5.72 from their allowances and earned $15.29 in their carwash fundraiser. How much did they have total?	Whole dollars _____ Quarters _____ Dimes _____ Nickels _____ Pennies _____

Addition with Decimals

Add to find the sums. Draw a line to connect to the right total.

$$\begin{array}{r} \$\ 315.73 \\ +\ \$\ 152.17 \\ \hline \end{array}$$

$ 298.78

$$\begin{array}{r} \$\ \ \ 71.48 \\ +\ \$\ 215.31 \\ \hline \end{array}$$

$ 467.90

$$\begin{array}{r} \$\ 127.56 \\ +\ \$\ 171.22 \\ \hline \end{array}$$

$ 368.16

$$\begin{array}{r} \$\ 337.04 \\ +\ \$\ \ \ 51.93 \\ \hline \end{array}$$

$ 286.79

$$\begin{array}{r} \$\ 160.10 \\ +\ \$\ 208.06 \\ \hline \end{array}$$

$ 388.97

God's love for _____ is greater than anything.

Decimal Addition

When adding with decimals, we first line up the decimal points. Then we write in a 0 where any numbers are missing.

The kids ran 3 miles on Monday, 1.5 miles on Tuesday, and 2.09 miles on Wednesday. How many miles did they run in all?

$$
\begin{array}{r}
3.00 \\
1.50 \\
+\ \underline{2.09} \\
6.59 \text{ miles}
\end{array}
$$

3.00 = 3 miles and no extra parts

1.50 = 1.5 miles and no hundredths

2.09 = 2.09 miles

Ada ran 0.8 miles on Monday and 0.39 miles on Tuesday. How many did she run in all?	On Monday, Les ran 1 mile. On Tuesday, he ran 0.55 miles. How many total miles did he run?	Violet ran 0.6 miles on Monday and 1.04 miles on Wednesday. How many miles did she run in all?

Decimal Addition Stories

Line up the decimals, then add to solve the addition stories. Rewrite the sums at the bottom of the page in order from least to greatest.

Jane cut 4.6 inches of ribbon for her doll's hat. Then she cut 11.2 inches for her dress. How many inches of ribbon did she cut in all?

The brothers spent 2.25 hours doing schoolwork and then 3.5 hours playing outside and building a fort. How many hours is this all together?

Jeremy drove 6.3 miles in the morning and then 9.75 miles in the afternoon. How many miles total did he drive today?

She drank an 8.8 ounce smoothie for breakfast. Then she ate 7.4 ounces of oatmeal later that morning. How many ounces total did she eat and drink?

_____   _____   _____   _____

least   greatest

The Lord listens to _____ and He cares.

Addition with Money Stories

Find the sums. Color in the correct choice of dollars and coins to make up each sum.

Alex has $7.83 and Andrea has $4.19 in their piggy banks. How much money do they have in total?

Sum = _____

11 dollars, 2 pennies	12 dollars, 2 pennies	12 dollars, 2 dimes, 1 penny

Marla has $2.10 and Martin has $3.80. How much money do they have?

Sum = _____

5 dollars, 3 quarters, 1 dime, 1 nickel	5 dollars, 3 quarters, 2 dimes	5 dollars, 3 quarters, 1 nickel

Cassie and Calvin each have $6.55. How much do they have all together?

Sum = _____

13 dollars, 1 nickel	12 dollars, 1 dime	13 dollars, 1 dime

Paul has $4.90 and Patty has $5.23 saved. What is their total amount?

Sum = _____

10 dollars, 1 dime, 3 pennies	10 dollars, 1 nickel, 3 pennies	9 dollars, 3 quarters, 2 dimes, 3 pennies

Decimal Addition Review

Add up the money amounts and find the sums.

John has $5.50 and he finds 3 quarters. How much money does he have now?

Jackie has $9.00 and she finds 2 dimes and 4 pennies. How much money does she have now?

Madison has $7.21 and she finds 1 quarter, 2 nickels, and a penny. How much money does she have now?

Melanie has $4.34 and she finds 1 dollar bill and 3 dimes. How much money does she have now?

Jacob has $8.13 and he finds 5 quarters and 9 pennies. How much money does he have now?

Max has $14.10 and he finds 6 nickels. How much money does he have now?

Week 26

Subtraction with Decimals

To subtract decimal numbers, first line up the decimals.
Then write in a 0 where any numbers are missing.

4.3 - 2.18

$$
\begin{array}{r}
4\,.\,3 \\
-\ 2\,.\,1\,8 \\
\end{array}
$$

$$
\begin{array}{r}
{}^{2\ 1} \\
4\,.\,\cancel{3}\,\cancel{0} \\
-\ 2\,.\,1\,8 \\
\hline
2\,.\,1\,2 \\
\end{array}
$$

Practice writing in the missing 0 in the hundredths column
before subtracting these decimal equations.

$\begin{array}{r}7.5\\-\ 5.23\end{array}$	$\begin{array}{r}1.7\\-\ 0.94\end{array}$	$\begin{array}{r}5.6\\-\ 3.38\end{array}$
$\begin{array}{r}2.9\\-\ 1.36\end{array}$	$\begin{array}{r}5.4\\-\ 4.15\end{array}$	$\begin{array}{r}9.8\\-\ 7.67\end{array}$
$\begin{array}{r}3.3\\-\ 0.29\end{array}$	$\begin{array}{r}8.9\\-\ 6.71\end{array}$	$\begin{array}{r}6.5\\-\ 3.42\end{array}$

Subtracting Decimals

Each set of subtracting decimal equations has one that is written incorrectly. Circle the one with the mistake and fix it.

4.5	4.54	4.54
-3.2	- 3.2	-3.2
1.3	4.22	1.34

6.05	6.05	6.05
- 4.3	-4.3	-4.30
5.62	1.75	1.75

7.92	7.92	7.92
-5.7	- 5.7	-5.70
2.22	7.35	2.22

9.70	9.7	9.70
-3.4	-3.4	- 3.4
6.30	6.3	9.36

5.8	5.80	5.80
-2.6	-2.6	- 2.6
3.2	3.20	5.54

8.6	8.60	8.60
-1.5	- 1.5	-1.5
7.1	8.45	7.10

God has saved _____ from the kingdom of this world.

Decimal Subtraction Stories

Math stories have clues to help you know what operation to do. These are words that tell us to use subtraction.

less	left over	less than
left	more than	difference

Underline the subtraction word clues in each story, then answer the questions.

The new slide at the park is going to be 8.2 feet long. The old one was only 6.4 feet long. What is the difference between the lengths?

The regular swings reach out 10.3 feet, but the toddler swings only reach 5.1 feet. How many more feet do the regular swings reach?

The playground fencing stretched a total of 273 yards. 25 yards of the fence fell down during a strong storm. How much of the fence was left standing?

The kids played outside on the playground yesterday and it was 86.4 degrees Fahrenheit. Today, the temperature only reached 81.8 degrees. How many more degrees was it yesterday than today?

Subtraction Stories

Notice the word clues in the sentences.
Write each decimal subtraction equation and solve.

On the weekdays, the park and playground is used for 2.5 hours total. On the weekends, it is used for 6.75 hours. How much more is it used on the weekend than on weekdays?	Mom told us we could play at the park for 3 hours total. We have been here for 1.75 hours. How much time do we have left?
For 2.2 years, the kids have enjoyed a sonic spinner on the playground. Sonic spinners only last 10 years. How much longer do they have to enjoy it?	His family lived near the park for 6.8 years. They moved and lived at their new house for 3.5 years. How much longer did they live at their old house than their new house?
The older kid playground was built 4.5 years ago, and the toddler playground was built 9.4 years ago. What is the difference in their ages?	The children are allowed to play on the bigger playground when they turn 8.5 years old. Malcolm is 7.75 years old. How much longer does he have to wait?

God has great plans for_____.

Subtraction with Money

When there is a whole number subtracting a decimal number, we write in a 0 for each decimal place that is missing.

We have 5 dollars and no coins. We spend $3.25.
How much is left?

$$\begin{array}{r} \$5 \\ -\ \$3.25 \end{array} \qquad \begin{array}{r} {}^{4}\ {}^{1} \\ \$\cancel{5}.00 \\ -\ \$3.25 \end{array} \qquad \begin{array}{r} \quad {}^{9} \\ {}^{4}\ \cancel{\cancel{1}}\ {}^{1} \\ \$\cancel{5}.00 \\ -\ \$3.25 \\ \hline \$1.75 \end{array}$$

All of these amounts only have dollars. They don't have any dimes, and they don't have any pennies. Practice writing in the decimal point and zeroes for both the dimes and pennies. The first one is done for you.

$96 = $96.00 $5 = $45 = $15 =

$70 = $11 = $20 = $2 =

$10 = $30 = $60 = $19 =

$50 = $25 = $83 = $0 =

Subtracting Money

Read the stories and write a subtraction equation for each.
Then solve it.

The parking fee for trucks and buses at the park is $4.80. The parking cost for cars is $2.35. How much higher is the cost for trucks than it is for cars?	The park earned $520.65 on a sunny day, but only $200.30 on a rainy day. What was the difference?
The money saved for the new playground is $1554.27. The swings cost $317.11. How much money is left?	A family drove in their car to the park and paid for parking. They paid with a $20 dollar bill. How much money did they have left?

God loves _____so much.

Decimals are Parts of a Dollar

We can use quarters to understand how decimals and fractions are parts of a whole.

50¢ = half of a dollar = 1/2   25¢ = 1 quarter = 1/4
75¢ = 3 quarters = 3/4   1 whole dollar = 4 quarters

$4.50 = 4 dollars and 5 dimes and 0 pennies

This is the same as writing 4 1/2 dollars.

Draw a line to connect the matching amounts.

3.75   3 1/2

3.25   3 1/10

3.5   3 3/4

3.00   3 0/4

3.10   3 1/4

Subtraction Stories

Read each story. Fill in the blank in each subtraction equation with the correct decimal number from the number bank at the bottom of the page. Then, answer the decimal subtraction problems.

There are 45 total hours of work needed to put in the new playground and fix all the bushes and flower beds. The crew has already worked for 25 1/4 hours. How many work hours do they have left?	45 - _____

The new grass and plants will cover a total of 16 acres of land. They finished covering 13 3/4 acres. How many more acres do they have left?	16 - _____

After finishing the planting, they started building the playground. Out of the total 22 different items, they have already put together 10 1/2 of them. How many playground items are left?	22 - _____

13.75	25.25	10.5

God loves to spend time with _____.

Decimal Subtraction Review

Answer each subtraction equation by circling the
correct answer out of the choices given.

When cleaning up the park, they found $13.28 left behind. The
next day, they found $5.37. How much less is this amount than
the first amount?

$7.11

$8.09

$7.91

On Saturday, the kids played on the slides for a total of 0.75
hours. They played on the uneven bars for 2.25 hours. How many
more hours did they use the uneven bars than the slides?

2.25

1.5

2.5

They played on the uneven bars on Sunday for 1.5 hours so far.
They were allowed to stay for 4.25 hours total. How much time
do they have left?

3.75

2.75

3.25

The new ropes for the climbing wall will cost $47.52. The total
budget was $100. How much money is leftover?

$53.52

$53.48

$52.48

Decimal Review

In the graph below, each slide is equal to 0.25 hours of time on the slides. Each set of swings is equal to 0.6 hours of time on the swings. Each ball is equal to 0.75 hours of time playing soccer. Answer the questions about the graph by using decimal addition and subtraction.

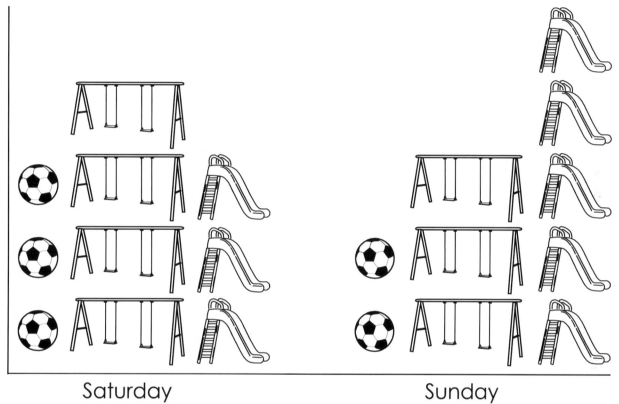

Saturday   Sunday

Which playground item was most popular this weekend?

How many more hours were spent on the swings than the slide on Sunday?

Which playground activity was played on the longest on Saturday?

What is the difference in hours between the most popular activity on Saturday and Sunday?

Unit 13
Week 27-28

Time

Unit 13 Instructions

Unit 13 reviews and expands on the concept of measuring time that was first introduced in Grade 1 math. Students review the quarter past, quarter to, half past, and o'clock measurements for minutes and learn the quarter hours. There is a lot of clock practice, both in reading a clock and drawing in the hour and minute hands for a given time. They also learn and practice the understanding of seconds, minutes, hours, days, weeks, months, and years.

The best way to practice and master the understanding of time is with their very own non-digital clock and calendar. Both of these items can be purchased very inexpensively and will be worth any investment. Students can practice moving the hands of the clock and even how to set an alarm. They can use a calendar to write in special dates, events, and holidays and practice reading the days and months of the year.

As with all lessons, parents can continue the learning process well past official school time by asking lots of questions and keeping the conversation going. "What is your favorite season/ day/month of the year? Why? How many days is it until (special day)? Can you look at your calendar and count it for me? How many weeks away is (special day)?" Have your student help plan the weekly calendar for the family and be the family secretary by writing in upcoming field trips or even a dinner meal plan.

Once they understand how to read a clock and calendar, parents can make the connections that minutes are parts or fractions of an hour, and months are parts or fractions of a year. Use your white board to show that 1 month is 1/12 of a year. 15 minutes is 1 quarter of an hour, which is like a money quarter, and can be written as 1/4 or 0.25.

Week 27

Time and Clock Review

You have learned a lot of facts
about telling time and reading clocks.

a.m. = **a**nte **m**eridiem.
This means "before the middle".
a.m. is all the times before the
middle of the day.

p.m. = **p**ost **m**eridiem.
This means "after the middle".
p.m. is all the times after the
middle of the day.

Let's review! Color as directed and rewrite the times below.

"Quarter past" is 15 minutes,
or 1 quarter, past the hour.
Color in the first quarter.

 7:15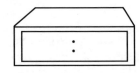

"Half past" is 30 minutes past
the hour. Color in the first
half of the hour.

 7:30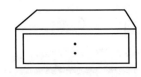

"Quarter to" is 15 minutes, or
1 quarter, before the hour.
Color in three quarters.

 7:45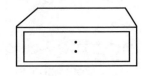

O'clock is exactly on the
hour. 60 minutes past
each hour. Color in all
4 quarters of the hour.

 8:00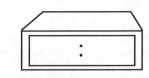

Clock Vocabulary Review

Draw a line to match up times and vocabulary.

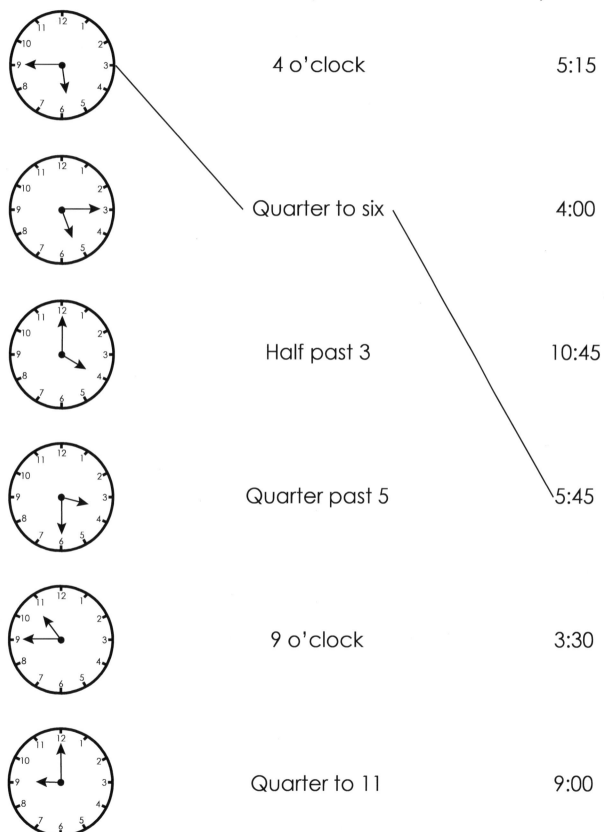

4 o'clock   5:15

Quarter to six   4:00

Half past 3   10:45

Quarter past 5   5:45

9 o'clock   3:30

Quarter to 11   9:00

God's favor rests on _____ .

Clock Times

Draw hour and minute hands on each clock
to match the times given.

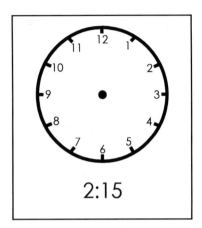

2:15

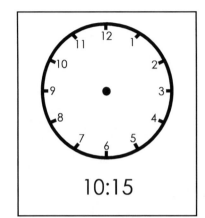

10:15

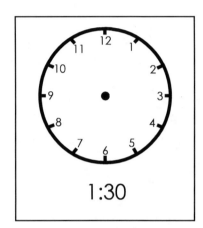

1:30

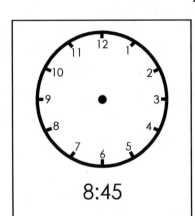

8:45

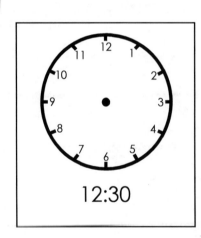

12:30

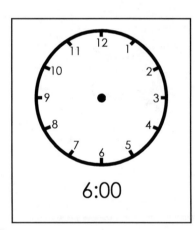

6:00

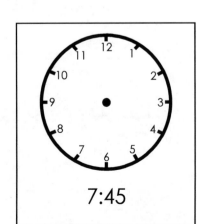

7:45

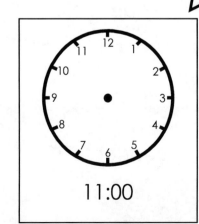

11:00

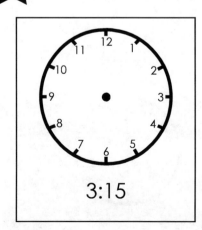

3:15

Reading a Clock

Look at the clocks below and circle the correct time for each.

   1:09   1:45   9:00

   9:06   6:45   9:30

   8:15   3:45   8:30

   12:15   3:12   3:00

   4:06   4:30   6:15

God speaks through _____ .

Periods of Time

The church service
started at 10:00 a.m.

Everyone left church
by 12:30 p.m.

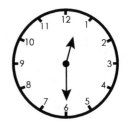

How much time passed
from start to finish?

First we look at the hours: 10 to 12 is 2 hours.
Then we look at the minutes: o'clock to 30 is 30 minutes.

The total time passed is 2 hours 30 minutes.

Write down the time shown on each pair of clocks.
Then solve how much time has passed from start to finish?

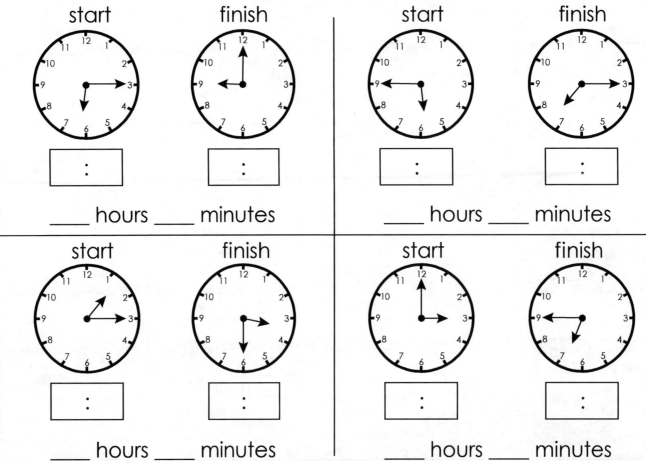

start   finish   start   finish

___ hours ___ minutes   ___ hours ___ minutes

start   finish   start   finish

___ hours ___ minutes   ___ hours ___ minutes

Comparing Time

In each set, circle the activity that takes the most time.

eating a sandwich   cooking dinner   driving to vacation

planting a flower   sneezing   sleeping all night

getting the mail   playing a board game   washing a dish

getting dressed   brushing teeth   taking a bath

   reading a   reading the   memorizing a
                        Bible verse   whole Bible   Bible verse

baking cookies   eating a cookie   buying cookies

coloring a picture   drawing a picture   showing a picture
                                            to Mom and Dad

God has great and infinite love for _____ .

AM and PM

A new day starts at midnight. All of the hours between the start of the day and the middle of the day are a.m. hours. The morning is a.m.	All of the hours between the middle of the day and the end of the day are P.M. hours. The afternoon and night are p.m.

Write in the time for each activity and circle a.m. or p.m. to show if it was in the morning, or in the afternoon and evening.

 brushing your hair _____ a.m. or p.m.

 eating dinner _____ a.m. or p.m.

 playing tennis _____ a.m. or p.m.

 working on math _____ a.m. or p.m.

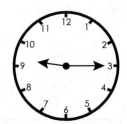 using a flashlight _____ a.m. or p.m.

AM and PM Activities

Rewrite the activities into the correct column.

dreaming   watching sunset   watching sunrise

listening to a bedtime story   waking up   breakfast

saying "Good morning"   putting on pajamas   yawning

saying "Good afternoon"   reading your Bible   sleeping

AM

_____

_____

_____

_____

_____

_____

PM

_____

_____

_____

_____

_____

_____

_____ has the mind of Christ.

Time Stories

The brothers started cleaning their room at 10:00 a.m. Everything was put away and neat by 10:45 a.m. How long did it take to clean their room?

_____ hours and _____ minutes

Dad left to walk their dog Blue at 7:45 a.m. He got back at 8:15 a.m. How long was their walk?

_____ hours and _____ minutes

Charlie started reading at 2:15 p.m. He finished and went to play outside at 3:30 p.m. How long did he read?

_____ hours and _____ minutes

The family left home to drive to their vacation at 9:30 a.m. They arrived at 1:30 p.m. How long was their drive?

_____ hours and _____ minutes

Janelle and her mom went shopping for a new dress. They left at 11:45 a.m. and were home by 2:15 p.m. How long were they gone?

_____ hours and _____ minutes

The kids' soccer game started at 4:00 p.m. The final whistle blew at 5:30 p.m. How long was the game?

_____ hours and _____ minutes

More Time Stories

Read each story and draw in the clock hands to show the start time and ending time. Write down the total time passed.

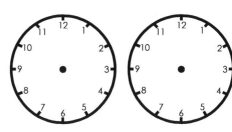

In the morning, Dad started our family devotions at 8:30. We finished at 9:00. What was the total time?

_____

Then we started our chores and getting ready for the day at 9:00. This took until 9:15. What was the total time?

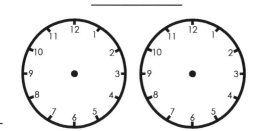

_____

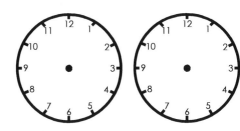

At 9:15 we started working on math and then history. We stopped for a break at 10:30. What was the total time?

_____

From 10:30 to 11:00 we walked the dog and played outside. What was the total time?

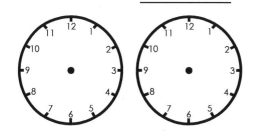

_____

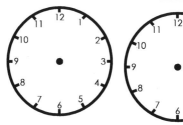

From 11 to 2 we ate lunch and finished our schoolwork for the day. What was the total time?

_____

From 2 until 5:30 we went on a field trip to the park and learned about animal habitats. What was the total time?

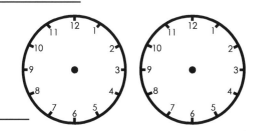

_____

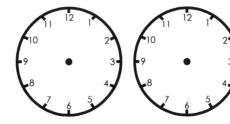

After dinner and before bed, we played games with the whole family from 7:30 - 8:45. What was the total time?

_____

Week 28

_____ is a joint heir with Christ.

Measuring Time
Minutes, Hours, and Days

24 hours is one full day.   48 hours is 2 days.

12 hours is 1/2 of a day   1 hour is 60 minutes

Time is always passing, except with the Lord. He lives outside of time and space. We can estimate how much time is passing when thinking about different activities we do during the day.

Draw a line to match up equal minutes, hours, and days.

24 hours   2 days

120 minutes   12 hours

90 minutes   1 hour

48 hours   1 day

1/2 day   36 hours

60 minutes   2 hours

1 1/2 days   1 1/2 hours

Passing Time

Rewrite these activities in order from the shortest amount of time to the longest amount of time.

Walking the dog _____

Watching a movie _____

Blinking your eyes _____

Making a sandwich _____

Doing a math lesson _____

Driving to Grandma's house _____

Summer vacation _____

Father God helps _____ learn many things.

Measuring Time – Weeks

7 days is equal to 1 week.	Sunday
	Monday
14 days is equal to 2 weeks.	Tuesday
	Wednesday
21 days is equal to 3 weeks.	Thursday
	Friday
	Saturday

Choose true or false for each statement below.

| There are 20 days in 2 weeks.

true/false | The weekend is 2 days: Sunday and Monday.

true/false | 2 weeks is 14 days.

true/false |
|---|---|---|
| 3 weeks is equal to 21 days.

true/false | Friday is the last weekday of the week.

true/false | The number of days in a week is an even number.

true/false |
| We can count the days that are in a number of weeks by skip-counting by 7s.

true/false | The first weekday is Sunday.

true/false | 10 days is greater than 1 week.

true/false |

Days of the Week

Solve the clues below about days and weeks.

After receiving a vision from God, Daniel fasted for 3 full weeks. How many days was this?

God created the whole earth and everything in it in only 1 week. How many days was this?

The people of Israel celebrated the Feast of Unleavened Bread for 2 whole weeks. How many days was this?

Jesus was tempted for 40 days in the wilderness. This is close to how many weeks?

On the 7th day of Creation, God rested. What is our 7th day of the week?

What is the middle day of the week?

Jonah was in the great fish for 3 days. Is this closer to 1/2 week or closer to 1 week?

The Lord watches over _____ .

Measuring Time - Months

"30 days has September, April, June, and November. All the rest have 31, except February which has 28, and 29 in a leap year."

Each season, winter, spring, summer, and fall, lasts 3 months.

winter: January, February, March

spring: April, May, June

summer: July, August, September

fall: October, November, December

Look at the amounts below and determine which is greater.
Use the >, <, and = symbols to compare amounts.

number of days in September		number of months in the fall		number of days in August		number of days in June
total number of months		number of days in January		number of days in the spring		number of months in the summer
total number of seasons		total number of months		number of Feb. days in a year		number of Feb. days in leap year
number of days in March		number of days in April		number of days in a month		number of weeks in a month

Calendar Fun

Use the practice calendar below to answer the questions.

March 2025

Sun.	Mon.	Tues.	Wed.	Thurs.	Fri.	Sat.
						1
2	3	4	5	6	7	8
9	10	11	12	13	14	15
16	17	18	19	20	21	22
23	24	25	26	27	28	29
30	31					

What day of the week is the first day of March? _____

What day of the week is the last day of March?_____

What day of the week is March 25th? _____

What day of the week will it be on April 1st? _____

How many total days are in March? _____

Jesus sticks closer than a brother to_____.

Measuring Time - Years

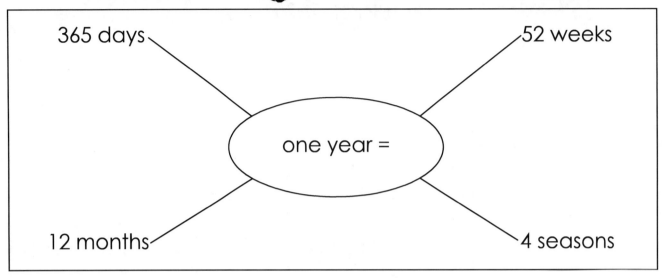

one year =
- 365 days
- 52 weeks
- 12 months
- 4 seasons

Color in 1/2 of the year:

12 months

Jan	Feb	Mar	Apr	May	Jun	Jul	Aug	Sep	Oct	Nov	Dec

4 seasons

Winter	Spring	Summer	Fall

Color in 1/4 of the year:

12 months

Jan	Feb	Mar	Apr	May	Jun	Jul	Aug	Sep	Oct	Nov	Dec

4 seasons

Winter	Spring	Summer	Fall

Fractional and Decimal Parts of a Year

Calculate the amount of time and fill in each blank.

I spend _____ months of the
3/4 of the year
year studying and learning math.

We have sunny and warm weather for

_____ months of the year.
0.5 of the year

Our family rakes leaves for

_____ months of the year.
1/4 of the year

We used _____ months of the
2 seasons of the year
year to build and plant a new garden.

They practiced tennis for _____
1/2 of the year
months before the tournament.

Our friends played on the baseball

team for _____ years.
24 months

Holy Spirit leads _____ into all truth.

Measuring Time Review

For each story, circle the best way to measure the time spent.

Our family went on vacation over Christmas break. hours years weeks	We dusted our rooms and changed our sheets. minutes months seasons
Grandma and Grandpa came with us to church this morning. days months hours	They moved to another country to work as missionaries. minutes years days
Everyone was outside raking leaves. weeks hours years	The kids played golf all season. days minutes months

Measuring Time Review

Use the word bank below to fill in the blanks.

The trees were planted many _____ ago. They grew

80 feet high. In the _____, the leaves drop off, and we

rake them. On _____, it took 2 _____ to finish the

yard work. Then, we spent 30 _____ playing in the huge

pile of leaves. The next day was _____, so we rested from

all the work. This process happens every fall _____. The fall

lasts for 3 _____, and then it is cold for the

3 _____ months. Soon, it is the month of _____, and

the leaves and the flowers start growing again. The

_____ are warmer, and we can wear shorts again.

Every year, we go through the same cycle of seasons for all

52 _____. God created hours, days, weeks, months,

seasons, and years with His great wisdom.

years	weeks	fall	Sunday
months	Saturday	minutes	winter
hours	days	season	April

Unit 14
Week 29-30

Measuring

Unit 14 Instructions

Unit 14 reviews and expands on the concept of measuring temperature and length as introduced in Grade 1 math. Your child will learn and practice reading a ruler and a thermometer. It will be very helpful to have a ruler, tape measure, and a thermometer on hand for them to practice these skills. You can hang an outside thermometer in the backyard and let them be the 'official checker' of the weather each morning. Ask them to give a weather report to the family and help decide what kind of clothes/jackets might be needed each day based on the temperature. You can make a pretend microphone with a paper towel tube and let them act as a meteorologist like on the news weather report.

If it's going to rain or snow, put out a bucket to catch all the precipitation. Once it stops (maybe the next day), have them measure the final amount with their ruler and write down the total inches and centimeters of rainfall or snow that fell in your yard. You can use a small thermometer and let them check the temperature of a glass of water. Show them the thermostat for your home and what temperature it is inside. Does this number change? Ask them why and when this might happen.

You can let them help measure the height of everyone in the family using the tape measure and mark these in a closet or garage wall. You can do this each month or quarter to track their height growth and write the inches next to each marking. Ask them what they want to measure, maybe it will be their foot or a favorite stuffed animal. The more they can practice using the ruler, tape measure, and thermometer, the more these math lessons will be reinforced, learned, and fully understood.

The Lord is good to _____ .

Measuring Temperature

We measure the temperature of the air around us and other things in degrees. It is a symbol that looks like a little circle.

26°C = 80°F

Most countries around the world use
the Celsius (C) scale of degrees, but the United States
uses the Fahrenheit (F) scale of degrees.

Color in each thermometer from the bottom up
to show the given temperature.

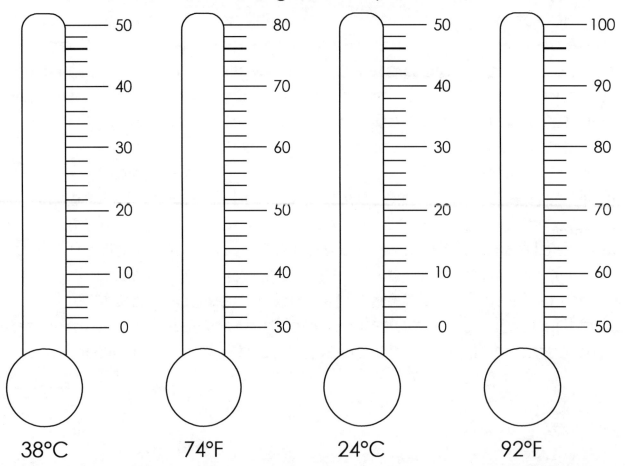

| 38°C | 74°F | 24°C | 92°F |

Celsius and Fahrenheit

Draw a line from each thermometer
to its matching temperature.

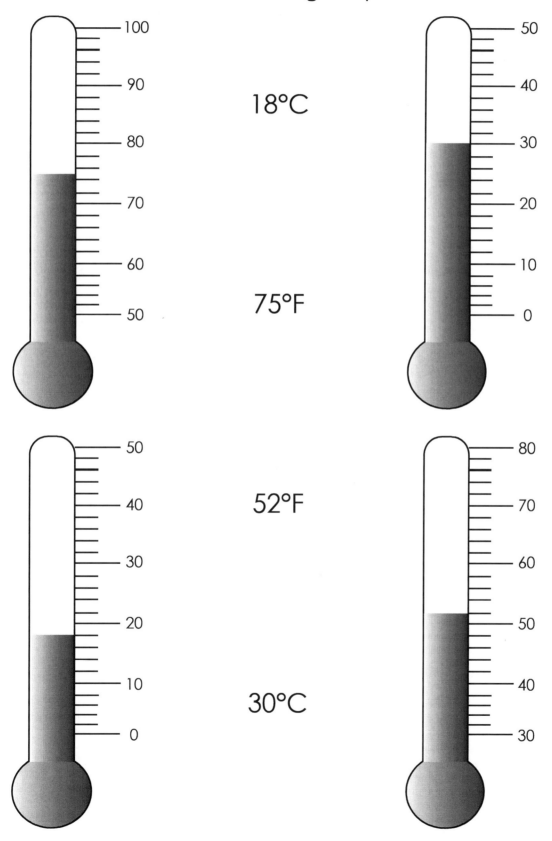

18°C

75°F

52°F

30°C

God is with _____ through every trial.

Boiling and Freezing Points

Water boils at:	Water freezes at:
100°C = 212°F	0°C = 32°F

Color in the thermometers
to show the boiling and freezing points of water.

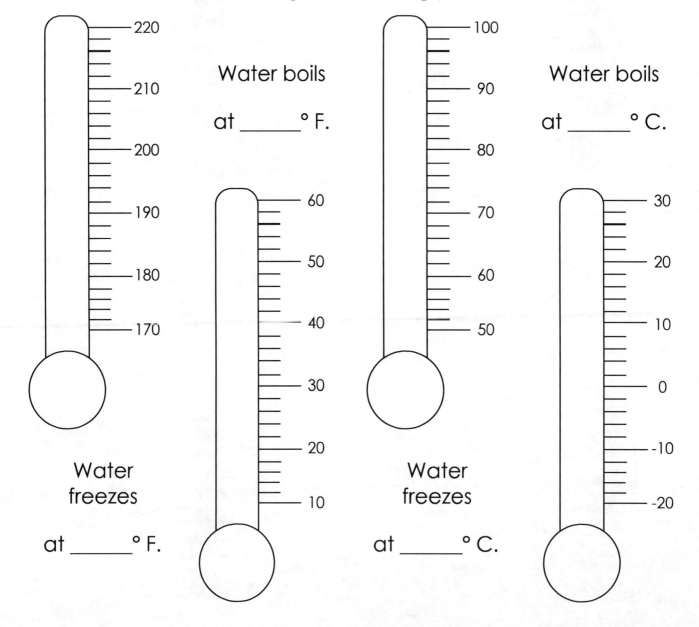

Water boils

at _____° F.

Water
freezes

at _____° F.

Water boils

at _____° C.

Water
freezes

at _____° C.

Boiling and Freezing

Each of the items below is at a temperature close to boiling or freezing. Write in the boiling point or freezing point.

_____ °F

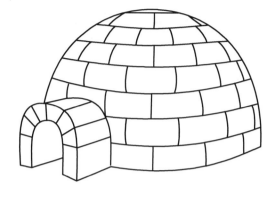

_____ °C

_____ °C

_____ °F

_____ °C

_____ °F

_____ is in love with Jesus.

A Week of Weather

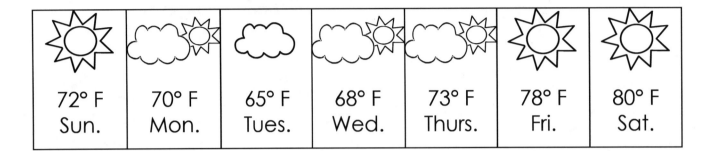

| 72° F Sun. | 70° F Mon. | 65° F Tues. | 68° F Wed. | 73° F Thurs. | 78° F Fri. | 80° F Sat. |

Study the weather calendar to answer these questions.

What is the highest temperature this week?

Day: _____ Temperature: _____

What is the lowest temperature this week?

Day: _____ Temperature: _____

How many days had full sun? _____

Which scale was used to measure the temperatures this week?

Fahrenheit or Celsius

Weather Continued

Put the temperatures from the previous page
in order from coldest to hottest.

_____ _____ _____ _____ _____ _____ _____

coldest   hottest

What is the difference in temperature between the hottest and the coldest day?

What is the difference between Monday's and Friday's temperatures?

What fraction of days out of the week were full sun?

What fraction of days out of the week had some clouds?

What is the temperature today where you live?

The Lord will never leave nor forsake _____.

Common Temperatures

The normal body temperature of a person is 98.6° F.
Anything higher is called a fever.

The temperature inside a home is around 70° F.

Any temperature below **zero** is a **negative** number.

Write the temperature that is showing on each thermometer.

_____ ° F

_____ ° C

_____ ° F

_____ ° F

Temperatures

Color in each thermometer to show the given temperature.

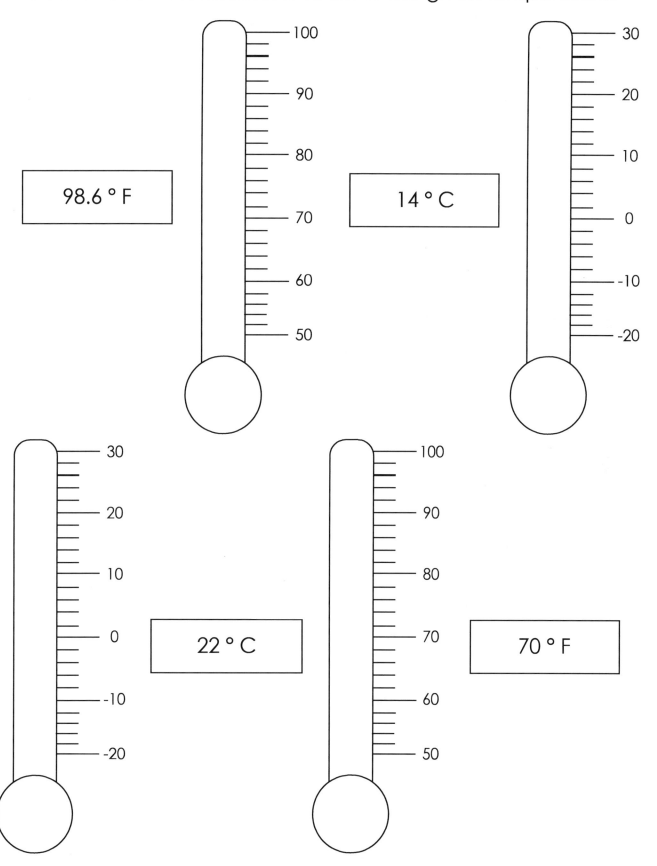

Jesus brings freedom to _____.

Temperature Review

Write the temperature shown on each pair of thermometers.
Color in the thermometer showing the higher temperature red
and the lower temperature blue. What is the difference
in degrees between them?

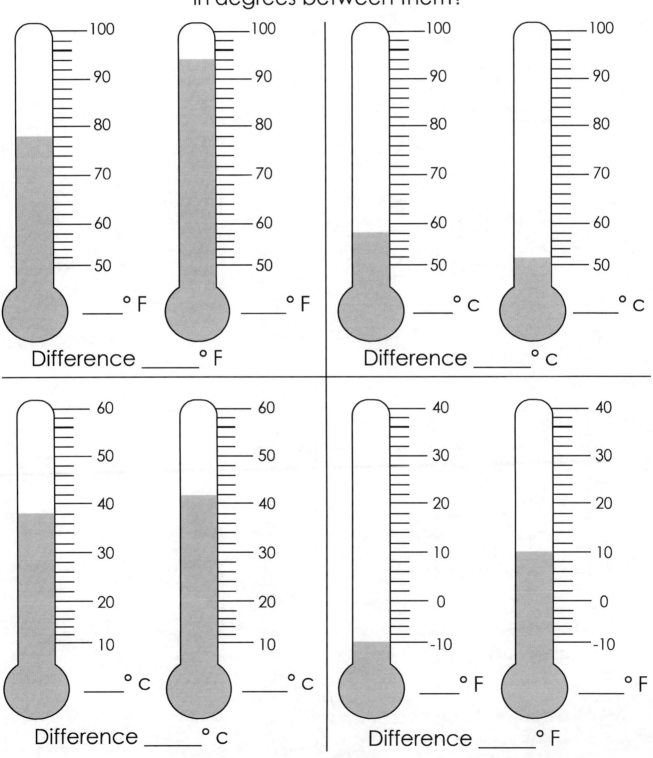

Reviewing Temperatures

Use the answer bank and fill in the blanks
to finish each sentence below.

The average temperature of a person is _____ ° F.

The boiling point of water is _____ ° C.

The freezing point of water is _____ ° C.

The average household temperature is _____ ° F.

The freezing point of water is _____ ° F.

The boiling point of water is _____ ° F.

70	0	98.6
100	212	32

Week 30

The Lord planned out a great future for _____ .

Measuring Length

The U.S. uses **inches**, **feet**, and **yards** to measure length and height. It is called the **Standard System**.

12 inches = 1 foot   3 feet = 1 yard

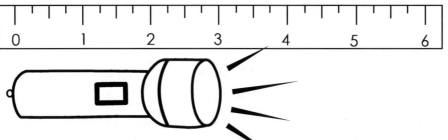

This mini flashlight measures 3 inches long.

1 inch

A quarter is about
1 inch across.

1 foot

This feather is about
1 foot long.

1 yard

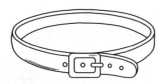

A man's belt is
about 1 yard long.

2 feet

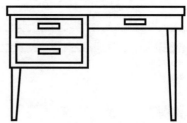

A desk is about
2 feet high.

How Long is it?

Using your own ruler, write down the length of each rope below in inches and in centimeters.

_____ inches   _____ centimeters

_____ inches   _____ centimeters

_____ inches   _____ centimeters

_____ inches   _____ centimeters

_____ inches   _____ centimeters

God's love surrounds _____ .

Metric System

Most other countries use **millimeters**, **centimeters**, and **meters** to measure length and height. It is called the **Metric System**.

10 millimeters = 1 centimeter   100 centimeters = 1 meter

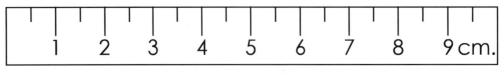

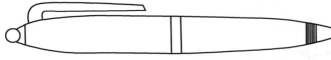

This pen is 7 cm. long.

Copy the measurements below into the blanks.

1 millimeter 1 mm.	1 centimeter = 1 cm.
_____	10 millimeters _____

The chain on a necklace is about 1 mm thick.

A bead is about 1 cm. across.

1 decimeter =   1 dm.

10 centimeters _____

1 meter =   1 m.

100 centimeters _____

The length of a popsicle stick is about 1dm.

The length of a guitar is about 1 m.

Metric Measurements

Write down the measurements of the shaded areas shown on each ruler below.

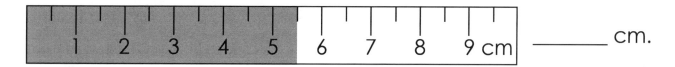 _____ cm.

 _____ cm.

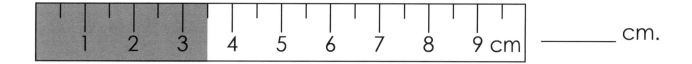 _____ cm.

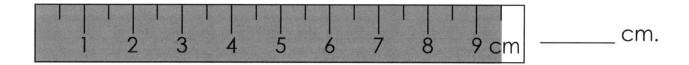 _____ cm.

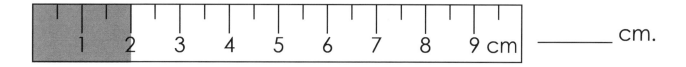 _____ cm.

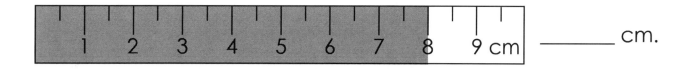 _____ cm.

God created _____ with gifts and talents.

Estimating Length

We can estimate the length of something
by using what we know about measurements.

Inches and **centimeters** are small, and we use them to measure things like a phone or a cup.

Yards and **meters** are big, and we use them to measure things like a football field or a building.

Feet are medium - sized, and we use feet to measure things like a room or a truck.

We use feet to measure the height of a camel.

Draw a line to match each item
to the correct unit of measurement.

inches/
centimeters

feet

yards/
meters

How Long is it?

For each picture, circle the correct measurement estimate.

8 inches   6 yards

9 feet   20 yards

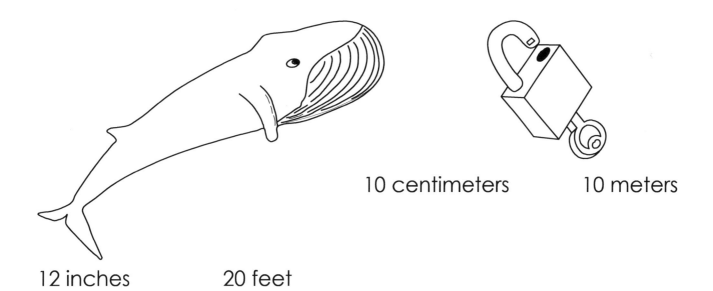

10 centimeters   10 meters

12 inches   20 feet

10 inches   100 yards

15 meters   15 millimeters

God's love for _____ cannot be measured.

Equal Measurements

Standard System

12 inches = 1 foot

3 feet = 1 yard

Metric System

10 millimeters = 1 centimeter

10 centimeters = 1 decimeter

100 centimeters = 1 meter

10 decimeters = 1 meter

Use the above measurements to fill in the blanks below.

24 inches = _____ feet

2 feet = _____ inches

3 feet = _____ yards

2 yards = _____ feet

36 inches = _____ feet

1 yard = _____ feet

20 millimeters = _____ centimeters

20 decimeters = _____ meters

30 centimeters = _____ decimeters

2 meters = _____ centimeters

3 centimeters = _____ millimeters

200 centimeters = _____ meters

Measurements

Read each statement below and circle if it is true or false.

2 yards is greater than 2 feet.	true /	false

20 centimeters = 2 decimeters	true /	false

6 feet = 3 yards	true /	false

36 inches is less than 2 feet	true /	false

2 decimeters = 20 centimeters	true /	false

40 millimeters = 3 centimeters	true /	false

4 meters = 400 centimeters	true /	false

_____ lives by faith in the Lord.

Measurement Review

Color in the squirrels below based on their measurement units.

red: inches

purple: yards

aqua: feet

green: centimeters

pink: millimeters

blue: meters

yellow: decimeters

 mm

 ft

 dm

 cm

 m

 in

 dm

 ft

 cm

 yd

 mm

 in

Length Review

Fill in the blanks of the crossword puzzle.

Across

1. _____ system uses m, cm, mm

3. 1 ft = 12 _____

6. 1 _____ = 10 millimeters

Down

1. 1 _____ = 100 cm

2. 1 _____ = 10 cm

4. 1 _____ = 3 feet

5. 3 _____ = 1 yard

Unit 15
Week 31-32

Measuring

Unit 15 Instructions

Unit 15 reviews and expands on the concept of measuring weight and capacity as introduced and practiced in Grade 1 math. Both standard and metric units of measure are introduced and reviewed to help gain full understanding.

Students are provided with a measurement chart and do *not* need to memorize measurement facts at this age. They can practice changing measurement amounts by using the equivalent measurement chart.

The best way to learn and grow in understanding is to practice with hands on activities. For this unit, helping in the kitchen and following recipes will be a great way to practice learning capacity. Using a measuring cup and pouring water to different levels will help with understanding. Pour 8 ounces of water into a bowl and into a cup and notice how the same amount looks different based on the size of the container holding it.

Students can practice weight by weighing themselves and weighing themselves holding various items to see any change in the amount.

Also in this unit is a review of subtraction, subtraction with borrowing, addition, addition with carrying, estimating, rounding to place value, and comparing values using greater than, less than, and equal to symbols. This extra practice with these previously learned lessons will increase proficiency.

The Lord heals _____ from any sickness.

Standard Weight

In the United States, we use standard weight measurements to show how light or heavy something is.

1 **ounce**	16 ounces = 1 **pound**	2000 pounds = 1 **ton**
1 oz	1 lb	1 t

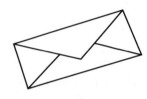  

Fill in the blanks below with the words:
1 ounce, 1 pound, or 1 ton.

A feather weighs about _____.

A giraffe weighs about _____.

A sweater weighs about _____.

A loaf of bread weighs about _____.

A dump truck weighs about _____.

A slice of bread weighs about _____.

Comparing Weight

Write the numbers 1, 2, and 3 under the items in each set
to show their weights, from the lightest (1) to the heaviest (3).

_____ has a calling from God.

Ounces and Pounds

We know from yesterday that 1 pound (lb) = 16 ounces (oz).

We can use this to change
ounces to pounds and pounds to ounces.

1/2 **pound** = 8 **ounces**
1/2 lb = 8 oz

2 **pounds** = 32 **ounces**
2 lbs = 32 oz

The pet store has a lot of food to deliver. Help them
by writing in the correct weights.

8 oz = _____ lbs	1 1/2 lbs = _____ oz

24 oz = _____ lbs	3 lbs = _____ oz

32 oz = _____ lbs	1/2 lb = _____ oz

16 oz = _____ lbs	2 lbs = _____ oz

Rounding Weights

Round each weight up or down to the closest pound.

31.6 lbs

_32_ lbs

12.5 lbs

____ lbs

15.1 lbs

____ lbs

11.9 lbs

____ lbs

8.4 lbs

____ lbs

1.7 lbs

____ lbs

20.2 lbs

____ lbs

2.6 lbs

____ lbs

10.8 lbs

____ lbs

25.3 lbs

____ lbs

The Metric System

Many countries around the world use the metric system to measure weight in grams and kilograms.

1 **gram** (g)   1000 grams = 1 **kilogram** (kg)

Change each amount from grams to kilograms.

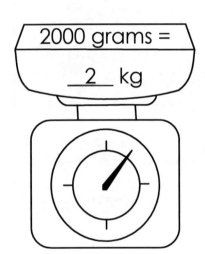
2000 grams = _2_ kg

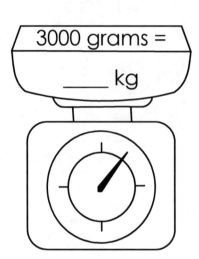
3000 grams = ___ kg

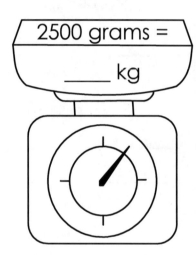
2500 grams = ___ kg

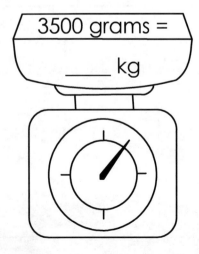
3500 grams = ___ kg

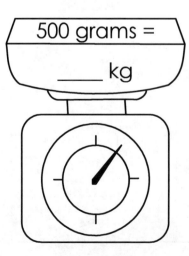
500 grams = ___ kg

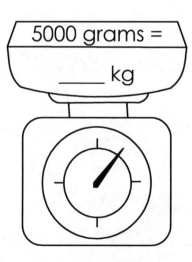
5000 grams = ___ kg

Adding Kilograms

Add up the kilogram weights.
How many grams is each amount?

2.5 kg
+ 1.0 kg
‾‾‾‾‾‾
3.5 kg

= 3,500 grams

7.6 kg
+ 2.4 kg
‾‾‾‾‾‾
___ kg

= _____ grams

4.5 kg
+ 4.5 kg
‾‾‾‾‾‾
___ kg

= _____ grams

15.8 kg
+ 2.2 kg
‾‾‾‾‾‾
___ kg

= ___ grams

1.3 kg
+ 9.2 kg
‾‾‾‾‾‾
___ kg

= _____ grams

3.4 kg
+ 5.1 kg
‾‾‾‾‾‾
___ kg

= _____ grams

1.3 kg
+ 8.7 kg
‾‾‾‾‾‾
___ kg

= ___ grams

9.0 kg
+ 4.5 kg
‾‾‾‾‾‾
___ kg

= _____ grams

4.2 kg
+ 2.8 kg
‾‾‾‾‾‾
___ kg

= _____ grams

0.2 kg
+ 4.3 kg
‾‾‾‾‾‾
___ kg

= ___ grams

2.1 kg
+ 5.4 kg
‾‾‾‾‾‾
___ kg

= _____ grams

11.6 kg
+ 3.4 kg
‾‾‾‾‾‾
___ kg

= _____ grams

_____ is a child of the One True and Living God.

Grams and Kilograms

Draw a line to connect the pictures with their matching weights.

90 kilograms

750 kilograms

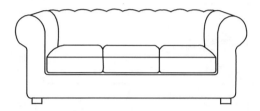

3 kilograms

10 grams

160 kilograms

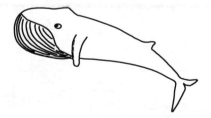

400,000 kilograms

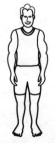

Review of Weight Measurements

Circle the correct word from the choices given.

The desk weighs 30 _____.

      pounds   ounces   tons

The house weighs 4 _____.

      kilograms   ounces   tons

The mouse weighs 20 _____.

      pounds   tons   grams

The piece of string weighs 3 _____.

      tons   kilograms   ounces

The full swimming pool weighs 6 _____.

      grams   tons   ounces

The Lord loves _____ unconditionally.

Weight Review

Compare weights by filling in each box with the symbols for
greater than >, less than <, or equal to =

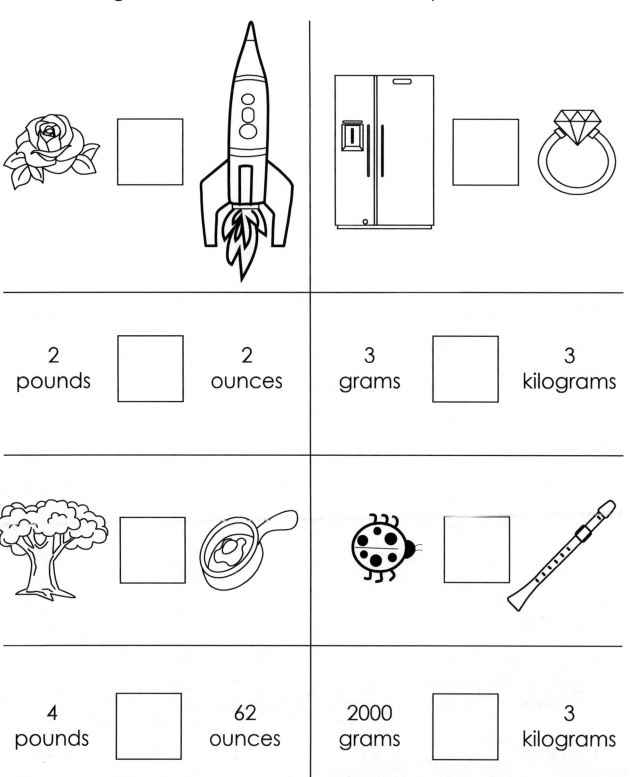

| 2 pounds | | 2 ounces | 3 grams | | 3 kilograms |

| 4 pounds | | 62 ounces | 2000 grams | | 3 kilograms |

Subtraction and Weights Review

Subtract each amount and write out
the full measurement word in your answers.

```
  1 5 . 7 lbs
-    9 . 2 lbs
  6 . 5 pounds
```

```
  8 . 8 g
-  1 . 4 g
```

```
  5 3 . 7 oz
- 2 4 . 3 oz
```

```
  3 . 9 kg
- 2 . 6 kg
```

```
  7 . 6 t
-    4 . 7 t
```

```
  5 . 5 lbs
- 3 . 7 lbs
```

```
  1 8 . 3 g
-    5 . 8 g
```

```
  2 1 . 4 oz
- 1 3 . 5 oz
```

Week 32

Holy Spirit empowers _____ to live a holy life.

Measuring Capacity

Capacity is the amount of space that something fills up.
We use standard capacity measurements for liquids.

1 fluid ounce	1 **cup** (c.)	1 **pint** (pt.)	1 **quart** (qt.)	1 **gallon** (gal)
(fl. oz.)	= 8 ounces	= 2 cups	= 2 pints	= 4 quarts

Write the numbers 1-3 to show least to greatest capacity
in each row of pictures.

_____ _____ _____

_____ _____ _____

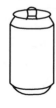  

_____ _____ _____

Ounces, Cups, Pints, Quarts, and Gallons

Write the correct and equal amounts in each blank.

The children drank _____ gallons of milk this week.
8 quarts

Mom mixed _____ pints of cocoa into the brownies.
4 cups

We put _____ quarts of gas into the lawn mower.
3 gallons

The lemonade container held _____ cups.
16 fluid ounces

The baby pool was filled with _____ quarts of water.
6 pints

Please put _____ pints of oil in the car.
1/2 quart

The casserole contained _____ fluid ounces of sauce.
1 cup

God is with _____ always.

Measuring Capacity

Look at each measuring cup and write down
the measurement of the liquids.

Measuring Capacity

Color in the correct picture for each amount.

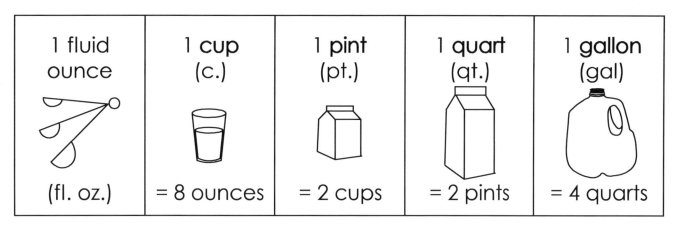

1 fluid ounce	1 **cup** (c.)	1 **pint** (pt.)	1 **quart** (qt.)	1 **gallon** (gal)
(fl. oz.)	= 8 ounces	= 2 cups	= 2 pints	= 4 quarts

1 quart

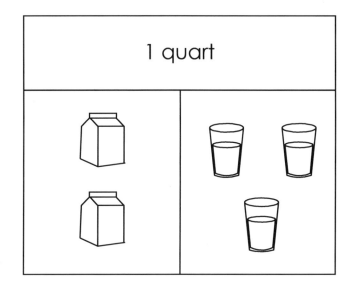

3 pints

16 fluid ounces

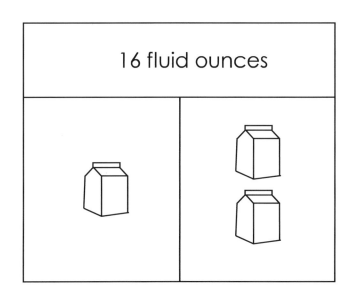

2 quarts

God loves _____.

Measuring Capacity with the Metric System

Sometimes, the metric system is used
to measure capacity and liquid amounts.

1 milliliter (ml)   1000 ml = 1 **liter (L)**   **2 liters**

 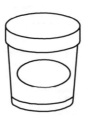 

Round each amount of milliliters up or down to the closest liter.

3490 ml → <u>3000 ml</u> = <u>3 liters</u>   5650 ml → _____ = _____

2843 ml → _____ = _____   9262 ml → _____ = _____

8370 ml → _____ = _____   6485 ml → _____ = _____

1199 ml → _____ = _____   7732 ml → _____ = _____

4910 ml → _____ = _____   3553 ml → _____ = _____

Milliliters and Liters

Circle the correct estimate amount for each picture.

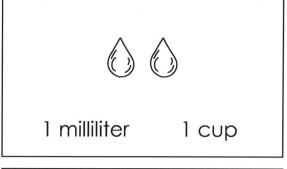

1 milliliter   1 cup

2 gallons   2 fluid ounces

4 pints   4 milliliters

6 quarts   6 gallons

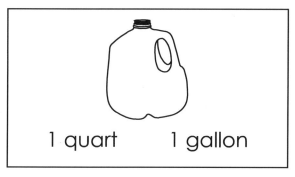

1 quart   1 gallon

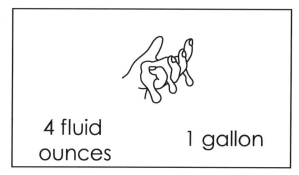

4 fluid ounces   1 gallon

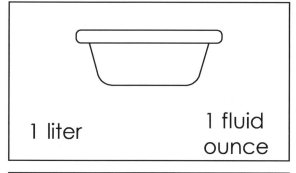

1 liter   1 fluid ounce

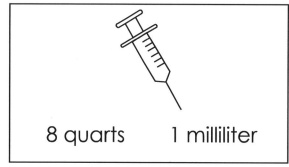

8 quarts   1 milliliter

1 milliliter   12 fluid ounces

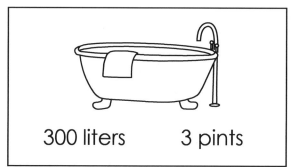

300 liters   3 pints

The Lord meets every need for _____ .

Adding Measurements

Add the measurements below.
Write in the correct abbreviations for each sum.

14 gallons + 26 gallons 40 g	29 quarts + 43 quarts	9 fluid ounces + 3 fluid ounces
270 pints + 143 pints	19 cups + 54 cups	276 liters + 316 liters
350 milliliters + 193 milliliters	514 gallons + 408 gallons	109 quarts + 851 quarts
426 liters + 553 liters	658 pints + 102 pints	135 cups + 712 cups

Subtracting Measurements

Subtract the measurements below.
Write in the correct abbreviations for each difference.

Phillip had a water bottle with 28 fluid ounces of water in it.
He drank 9 fluid ounces. How many were left?

Andrea filled the pool with 1120 gallons of water. Last summer,
it had 968 gallons in it. How many more does it have now?

Mom needs 8 cups of shredded cheese to make a casserole.
She has a bag with 64 cups in it. How many are left?

The store had 135 liters of soda for sale. They sold
16 liters today. How many are left?

The home school graduation party will serve 75 pints of ice
cream to the guests. There are 22 pints in the freezer now.
How many pints do they still need to buy?

_____ follows the leading of Holy Spirit.

Capacity Measurement Review

Circle whether each statement is true or false.

5400 milliliters = 5 liters and 400 ml	True / False

1298 milliliters = 12 liters and 98 ml	True / False

2050 milliliters = 20 liters and 50 ml	True / False

3612 milliliters = 3 liters and 612 ml	True / False

4481 milliliters = 4 liters and 481 ml	True / False

7338 milliliters = 7 liters and 338 ml	True / False

6529 milliliters = 65 liters and 29 ml	True / False

Capacity Review

Draw a line to connect the matching amounts.

8 cups	3 cups
2000 milliliters	2 pints
1 quart	2 quarts
1 gallon	6 quarts
24 fluid ounces	16 cups
12 pints	4000 milliliters
4 liters	2 liters

Unit 16
Week 33-34

Graphs and grids

Unit 16 Instructions

Unit 16 introduces and provides practice in reading graphs, maps, and grids. Students will learn how to read and create a bar graph, line graph, pie chart, pictograph, and grid. They also learn how to read and answer questions about a variety of maps. An introduction is made to North, South, East, and West directions, a compass rose, a map key, and a map legend.

There are many opportunities for your student to practice creating their own graphs. They can extend this unit and have extra practice by thinking of their own questions and gathering data from family and friends.

A fun way to practice with direction and maps is to let your child practice with a map of your city and state. Have them locate the map key, legend, and compass rose on it. Ask them to locate your city or state on a map of the United States. Ask them to show you different things on the map such as rivers, oceans, borders, and landmarks. You can also have them draw a map of your house and property and label all the different rooms.

When working with direction, you can use an inexpensive compass and show them how to orient it to true north - it will come with an instruction booklet on how to do this. As you are walking or playing or driving together, they can use this to help determine in what direction you are traveling.

As with any math concept, practice will always result in proficiency. Over time, and with a lot of practice, your child will be able to understand maps and the wonderful subject of cartography. Along with this unit, you can find plenty of books in the library about various explorers that explain how they used the stars to determine direction and find their way to new lands.

Week 33

Line Graphs

We use **charts** and **graphs** to show information. Since they are pictures, they help us understand more quickly and easily.

Silas spent time each day practicing piano. Here are his times.

Sunday 0.5 hours	
Monday 2 hours	
Tuesday 1.5 hours	
Wednesday 1 hour	
Thursday 2 hours	
Friday 1 hour	
Saturday 0 hours	

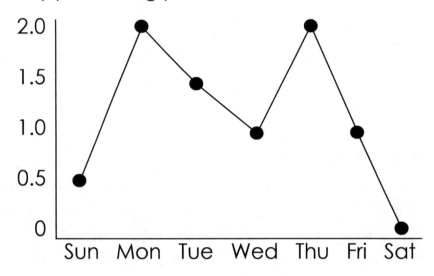

The picture above is called a **line graph**.

Which was easier and faster to read and understand: the list of hours on the left, or the picture on the right?

_____

On which day did Silas practice the least?

_____

On which day did Silas practice the most?

_____

Did he usually practice more on the weekdays, or on the weekends?

_____

Reading a Line Graph

Use the information given to draw a dot for each day's amount. Then, connect the dots to complete your line graph. Use your line graph to answer the questions below.

Silas's sister, Jackie, loves to read. Here is how much she read this week.

Sunday 2 hours
Monday 2.5 hours
Tuesday 1.5 hours
Wednesday 3 hours

Thursday 2 hours
Friday 2 hours
Saturday 1 hour

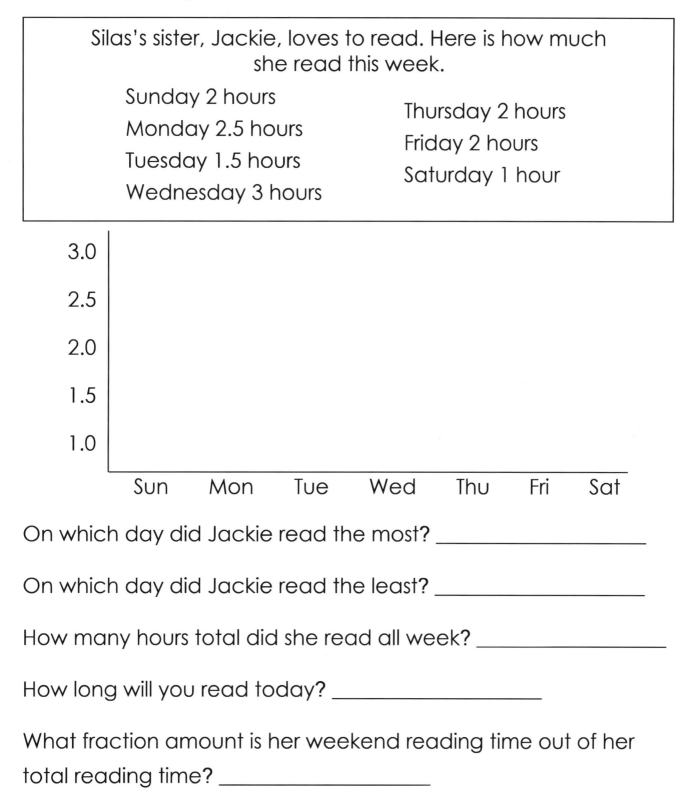

3.0

2.5

2.0

1.5

1.0

Sun   Mon   Tue   Wed   Thu   Fri   Sat

On which day did Jackie read the most? _____

On which day did Jackie read the least? _____

How many hours total did she read all week? _____

How long will you read today? _____

What fraction amount is her weekend reading time out of her total reading time? _____

Pie Charts

Another way to show information is with a **pie chart**, which is a circle divided into different sections.

Silas received $100 from his friends and grandparents friends for his birthday! Here is what he did with the money:

piggy bank - $60

bought a new building set - $25

gave to church - $10

bought a gift for Jackie - $5

Color in each section of the pie chart in a different light color so you can still read the words in it. Then, answer the questions.

Which is easier to understand: the list of his spending on the left, or the pie chart on the right?

_____

What did he do with most of his birthday money?

_____

What did he spend the smallest amount on?

_____

What fraction of the pie chart represents how much he spent on his new building set?

_____

Making a Pie Chart

Use the information given to write the correct label on each section. Color them in lightly with a different color for each piece of the pie. Use your pie chart to answer the questions below.

Jackie helped Mom do the grocery shopping this week. They spent $120 total.

Fruit	$22.50
Vegetables	$15
Desserts	$7.50
Meat/Chicken	$37.50
Breads/Bagels	$15
Drinks	$7.50
Dairy Items	$15

What did they spend the most money on?

_____

What did they spend the least amount of money on?

_____

How much more did they spend on meat than on vegetables?

_____

What fraction shows the amount of money spent on vegetables out of the entire amount? Use the pie chart picture to help you figure out the fraction.

_____

_____ casts all cares onto Jesus.

Bar Graphs

Another way to show information is with a **bar graph**.

Silas and Jackie saw that a number of people came to church Sunday mornings, and a different number of people came to the weekly prayer meetings.		
Week	**Sun.**	**Prayer**
May 4	135	76
May 11	190	80
May 18	151	36
May 25	98	48

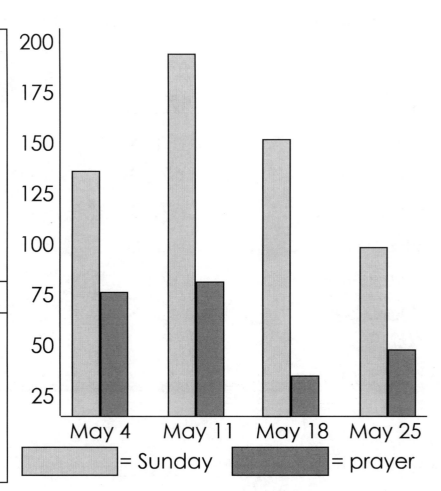

Which week had the highest attendance at Sunday service?

_____

Which week had the highest attendance at prayer meetings?

_____

Did more people attend on Sundays or prayer meetings?

_____

Bar Graphs

Use the information given to continue drawing each line to make a rectangle to show how many men and women went to church on Sunday. Color the men's rectangles one color and the women another color. Use your bar graph to answer the questions below.

week	men	women
Here are the numbers of men and women who attended Sunday morning church services.		
May 4	60	75
May 11	91	99
May 18	81	70
May 25	43	55

100
90
80
70
60
50
40

|m| |w| |m| |w| |m| |w| |m| |w|
May 4   May 11   May 18   May 25

[ ] men   [ ] women

Which week had the highest attendance overall?

_____

Which week had the highest attendance for women?

_____

On which week were there more men at church than women?

_____

Which week had the lowest attendance overall?

_____

God is a loving Father to _____ .

Pictographs

Another way to show information is with a **pictograph**.

After their busy week, the family went to the zoo over the weekend. They saw a lot of animals.

monkeys	10
lions	4
ostriches	12
bears	8
elephants	3

monkeys

lions

ostriches

bears

elephants

Each picture = 2 animals

Which type of animal did they see they most?

_____

Which type of animal did they see the least?

_____

Why do you think there is a picture of half an elephant?

_____

How many more ostriches did they see than elephants?

_____

How many total animals did they see?

_____

Line Graph Review

Now, you will get a chance to make your own graphs and charts from what you learned this week.

Line graph: Pick 1 thing you do every day, like schoolwork, playing outside, or playing with your favorite toy. Keep track for 1 week to show how many hours you spend on that activity. Then, draw a dot for each day's hours and connect the dots. Write a title at the top of your line graph to show what it is about.

_____

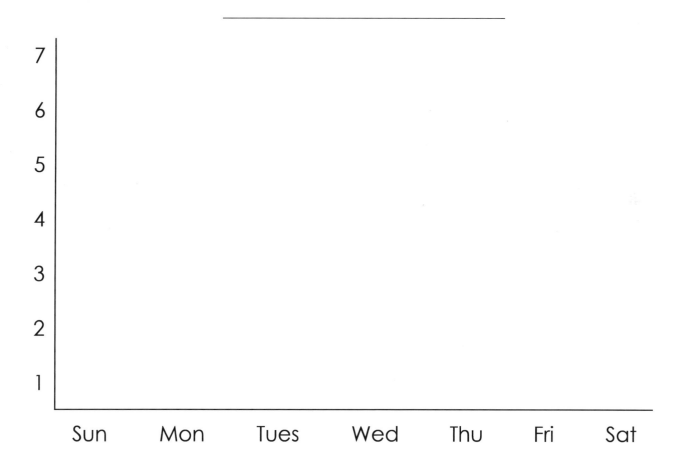

_____ is the apple of God's eye.

Pie Chart Review

Pie chart: Track all your activities for an entire day. This would include sleeping, eating, cleaning, school, playing outside, reading, and anything else you do. Color in the pie pieces in the chart below to show how your day was split up. Each section is equal to 3 hours, and you can split a section in half if you need to show less than 3 hours.

Write the activity that each section shows onto each colored-in section with a marker. Write a title like "My Day" or "All About Me" at the top of your pie chart to show what it is all about.

_____

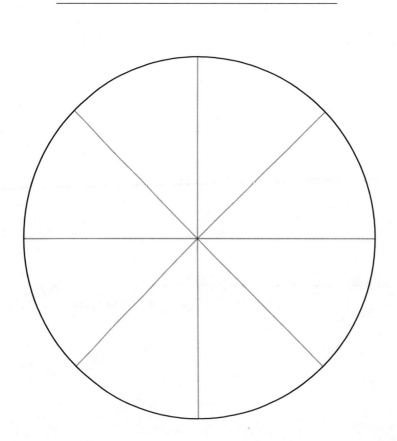

Bar Graph Review

Bar graph: Pick a question that you are curious about. Ask your mom or dad to post it on their social media and tell everyone that you are making a graph. Give them 4 choices to pick from. As the answers come in, you can keep track of them using tally marks. After you get enough responses, you can put all of the information into a graph below. Here are some ideas for questions, or you can make your own. Write a title at the top. Write in the 4 choices at the bottom and draw a rectangle bar to show the amount of answers for each choice.

What is your favorite ice cream flavor?
What is your favorite season?   What is your favorite subject?
What is your favorite pet?   What is your favorite food?

Week 34

_____ is a treasure to God.

Parts of a Map

Every map has different parts and colors that help us
to understand what it means.

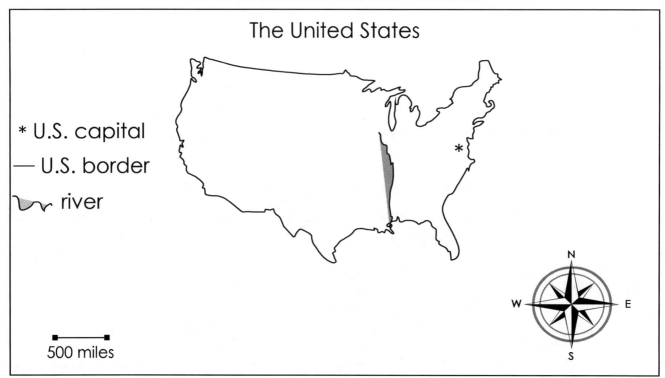

The **title** of a map shows us what the map is a picture of. What is
the title of this map? _____

The **scale** shows us the true size of the space shown on the map.
This map has a scale where each section measuring this size is
really _____ miles long.

The **compass rose** shows us which way on the map is north,
south, east, and west. Which direction on the U.S. map are you
heading if you go to the left? _____

The **key** shows us a list of all the types of things on the map and
what each one looks like. On this U.S. map, what does the star
mean? _____

Where is it Located?

Study the map and use it to fill in the blanks below.

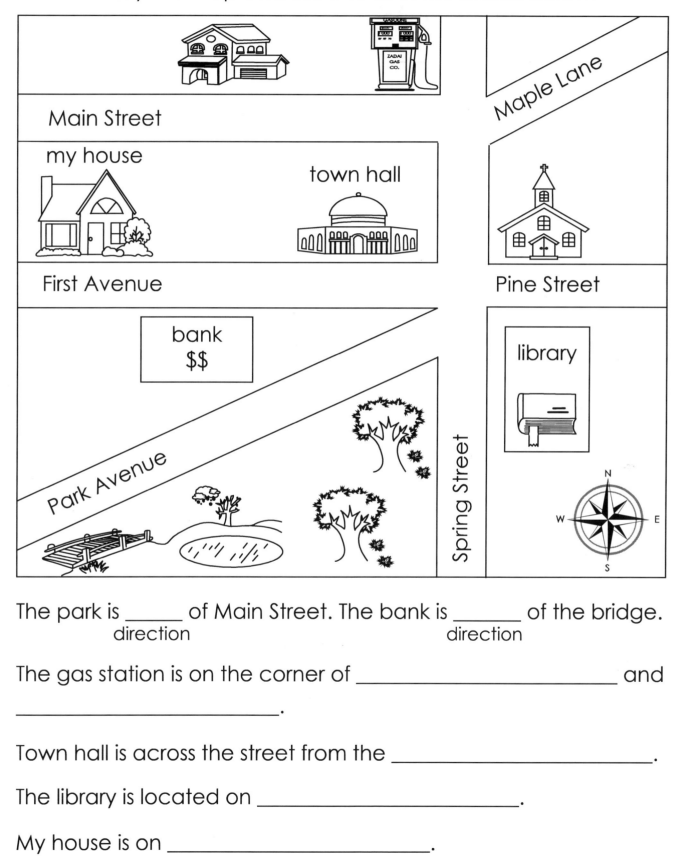

The park is _____ of Main Street. The bank is _____ of the bridge.
         direction   direction

The gas station is on the corner of _____ and

_____.

Town hall is across the street from the _____.

The library is located on _____.

My house is on _____.

The Lord loves _____ for eternity.

Traveling the World

Use the world map, compass rose, and scale
to fill in the blanks.

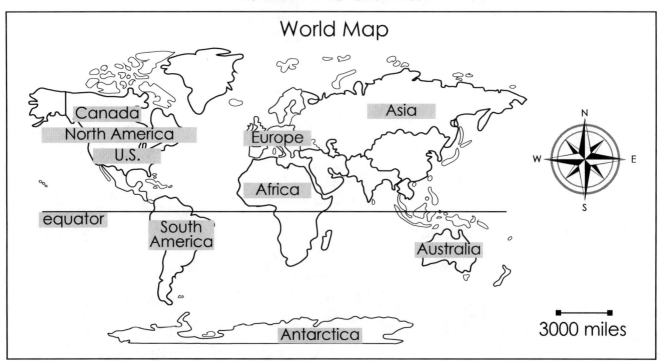

Daniela starts in the United States on the WEST coast. She flies
3000 miles EAST. Then she drives NORTH 500 miles.
What country is she in now? _____

She now flies EAST over a big ocean and sees land. She gets off
the plane and is in what continent? _____

Daniela eats a delicious Italian meal and then takes a boat
and a bus SOUTH to a very hot area closer to the Equator. What
continent is she in now? _____

She drives SOUTH to the very tip of the continent and then sails
EAST on a ship to an island continent.
Where is she now?   _____

Where would you like to go someday? _____

Making a Map

Follow the instructions to add to the town map.

1st Street

Lenny's house

grocery store

Purple Place

2nd Street

Lemon Avenue

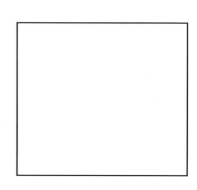

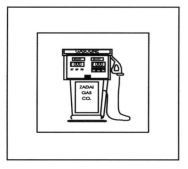

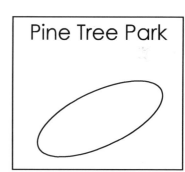
Pine Tree Park

Draw 2 trees at Pine Tree Park. Color the lake in blue.

Color all of Purple Place purple.

Color all of Lemon Avenue yellow.

Draw a library west of Lenny's house.

Draw a fire station east of the church.

Draw a bank in between the gas station and the park.

Draw a fountain in between the gas station and the library.

Draw a car driving east from the library to the grocery store.

Jesus gave His life for _____ .

Reading a Grid

A **grid** is a set of horizontal and vertical lines labeled with letters and numbers. We use these numbers and letters to pinpoint locations.

All the friends were at the playground today!

Hope is at G,3. Color her red.
Evan is at B,7. Color him green.
Matthew is at E6. Color him yellow.

Draw yourself into the playground. Where are you? _____ .

School Supply Grid

Use the grid below to answer the questions.

The kids found all their school supplies.

Where is the paper clip? _____   Where are the scissors? _____

Where is the pencil? _____   Where is the eraser? _____

Where is the calendar? _____   Where is the paper? _____

Where is the crayon? _____   Where is the glue? _____

Jesus came for _____ to have an abundant life.

Birthday Party Grid

Use the grid below to write in the location of each item.
My Birthday Party

birthday party hat _____   ice cream   _____

cupcake   _____   kazoo   _____

birthday crown _____   balloon   _____

Making a Grid

Draw the animals and other items listed
into the right areas of the farm.

On the Farm

	A	B	C	D	E	F	G	H
8								
7								
6								
5								
4								
3								
2								
1								

Draw a pig into E,5.   Draw a chicken into H,8.

Draw a horse into A,2.   Draw a sheep into C,7.

Draw a dog into F,4.   Draw a barn into G,1.

Draw any other animal or farm item into another section.

What did you draw? _____ Where is it located? _____

God is the author of _____'s life.

Review of Maps and Grids

Fill in the blanks using the word bank to finish each sentence.

The _____ on a map shows direction.

A plane flying from Europe to U.S. is going _____.

The _____ on a grid show left to right locations.

The _____ on a grid show up and down locations.

The map _____ shows what each item means.

The _____ on a map or grid shows what it is all about.

A boat sailing from Africa to Australia is traveling _____.

A map _____ shows how big or small everything is.

scale   numbers   west

key   letters

compass   east   title

Jesus and His Disciples Grid

Fill in the blanks using the word bank to finish each sentence.

Color in the disciple located at I,6 in pink.

Color in the disciple located at K,3 in blue.

Color in the disciple located at E,6 in green.

Color in the disciple located at K,7 in purple.

Color in the disciple located at I,4 in yellow.

Color in the disciple located at B,6 in gray.

Color in the disciple located at D,4 in brown.

Color in the disciple located at H,3 in red.

Color in the disciple located at J,6 in aqua.

Color in the disciple located at G,4 in black.

Who is standing at G,7? _____.

Unit 17
Week 35-36

Basic Geometry

Unit 17 Instructions

Unit 17 introduces and provides practice with both flat 2-dimensional shapes, as well as 3- dimensional solids. There is a lot of overlap with shapes and their definitions. For example, a square also fits the definition of a quadrilateral, a rhombus, and a parallelogram. It is just a more specific type. Hands on practice as usual will make learning math easier for many students.

One way to learn about shapes and solids is to have your child find objects in your house that are of each type. For a trapezoid, parallelogram, or rhombus, you can always draw and cut out shapes from construction paper. Then copy down the definitions given for each shape and solid. Lay these out on a table in separate areas or groups. One at a time, take an object and find out how many groups it fits into.

For example, let's say you have a square item. This would be able to join the quadrilateral group, the parallelogram group, the rhombus group, and the square group. It fits all of the definitions of those 4 shapes. You can continue on using all of the different objects they collected.

There is no need at this age to necessarily memorize all of the characteristics and overlap of every shape and solid. If they have a basic understanding of shapes and their characteristics, they can build on this foundation as they grow older.

You can also go on a scavenger hunt, either inside or outside, and work together to find and point out things that are shaped like triangles, circles, or rectangles.

The only math calculation used in this unit with shapes and solids is perimeter. At this age, they can learn how to measure the distance of the outside edge of a flat object using a ruler.

God forgives _____ of every sin.

Quadrilaterals - Parallelograms

A **quadrilateral** is any shape with four sides. When they have two sets of parallel sides, they are **parallelograms**.
Here are two special parallelograms:

4

4   4

4

square
every side is the same length
4 corners are all the same

12

6   6

12

rectangle
parallel sides are the same length
4 corners are all the same

Color in all of the squares and rectangles in blue.
Color in all the other shapes green.

Perimeter

The **perimeter** is the outside of a quadrilateral. Add up all 4 sides to find the perimeter for each square and rectangle below.

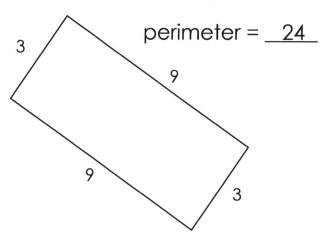

perimeter = __24__

perimeter = ____

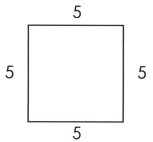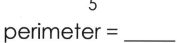

perimeter = ____

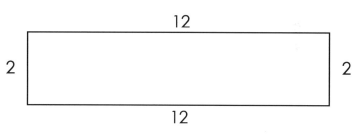

perimeter = ____

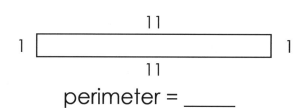

perimeter = ____

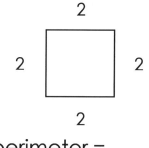

perimeter = ____

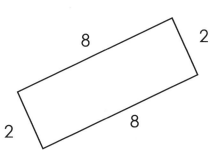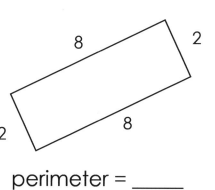

perimeter = ____

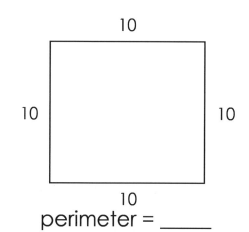

perimeter = ____

More Quadrilaterals – Parallelograms

A quadrilateral is any shape with 4 sides.
When they have 2 sets of parallel sides, they are parallelograms.

Here is another special parallelogram.

rhombus
every side is the same size.

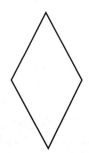

If it's not a rhombus, square, or rectangle, we call it a **parallelogram**.

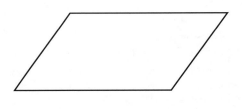

Use the clues to name each shape and draw an example.

I have 4 equal sides and look like a diamond. What am I? _____	

I have 2 sets of matching sides and 4 matching corners. What am I? _____	

I have 4 equal sides and 4 matching corners. What am I? _____	

I have parallel sides but am not a square, rhombus, or rectangle. What am I? _____	

Parallelograms and Symmetry

Many shapes have symmetry. They have 2 matching halves.
We can draw a **line of symmetry** to show the 2 equal sides.

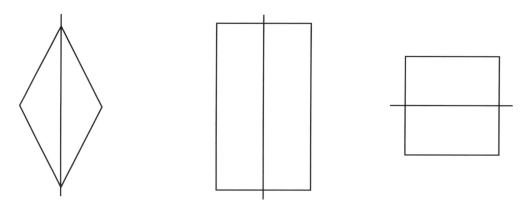

Draw a line of symmetry into each shape below
to divide into 2 equal halves. It may be a horizontal,
vertical, or diagonal line.

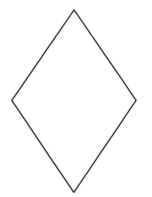
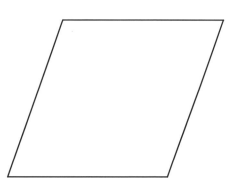
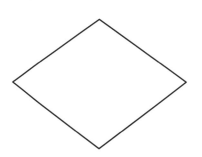

The Lord cares deeply for _____.

Quadrilaterals - Trapezoids and Kites

Some quadrilaterals do not have 2 sets of parallel sides.

A **trapezoid** has
1 set of parallel sides.

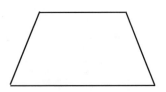

A **kite** has no parallel sides.
It has 2 sets of equal sides.

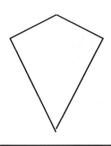

For each trapezoid below, highlight the set of parallel lines. If it is not a trapezoid, write a **K** for kite on the inside of the shape.

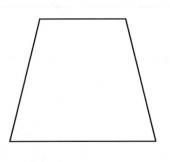

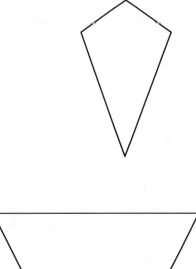

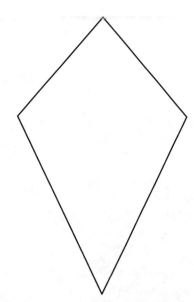

Quadrilaterals Review

Name each quadrilateral below.

God listens to _____, and it brings Him joy.

Polygons

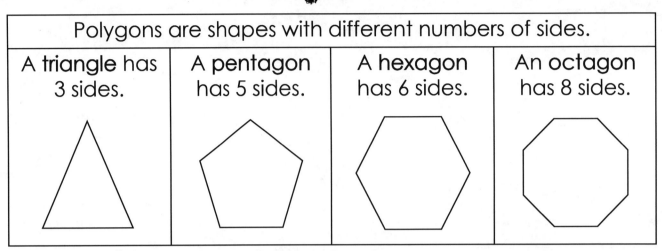

Polygons are shapes with different numbers of sides.			
A **triangle** has 3 sides.	A **pentagon** has 5 sides.	A **hexagon** has 6 sides.	An **octagon** has 8 sides.

Finish drawing each half of the polygons below across the lines of symmetry.

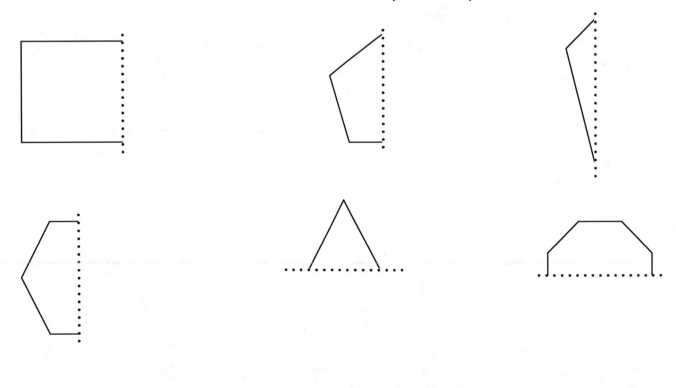

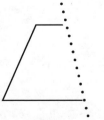

Polygons Review

Color in the polygons below using the following clues.

Color in the smallest triangle pink.

Color in the largest triangle green.

Color in the largest pentagon blue.

Color in the smallest pentagon purple.

Color in the smallest hexagon yellow.

Color in the largest hexagon gray.

Color in the smallest octagon orange.

Color in the largest octagon red.

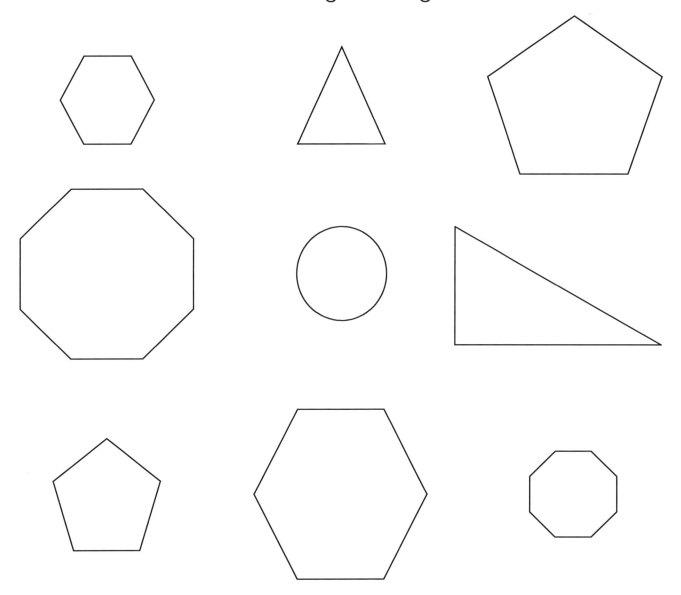

God speaks to _____ through His Word.

Types of Triangles

There are several types of triangles, and they all have 3 sides.			
An **equilateral triangle** has 3 equal sides.	An **isosceles triangle** has 2 equal sides.	A **scalene triangle** has 0 equal sides.	A **right triangle** has 1 right angle that equals 90°.

Circle the correct name for each triangle listed below.

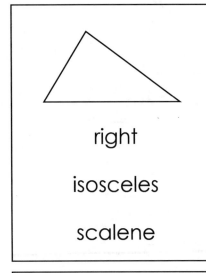

right

isosceles

scalene

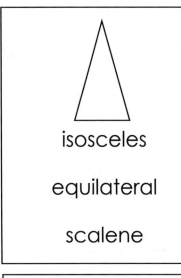

isosceles

equilateral

scalene

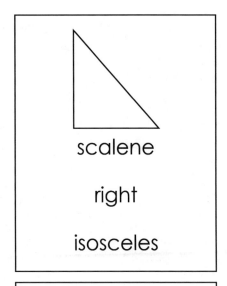

scalene

right

isosceles

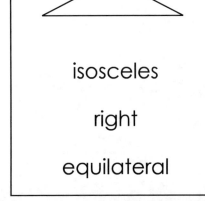

isosceles

right

equilateral

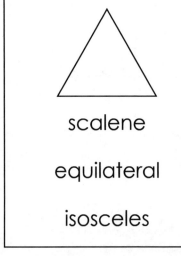

scalene

equilateral

isosceles

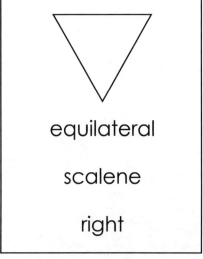

equilateral

scalene

right

Shapes Review

Draw a line to connect the matching shapes and their names.

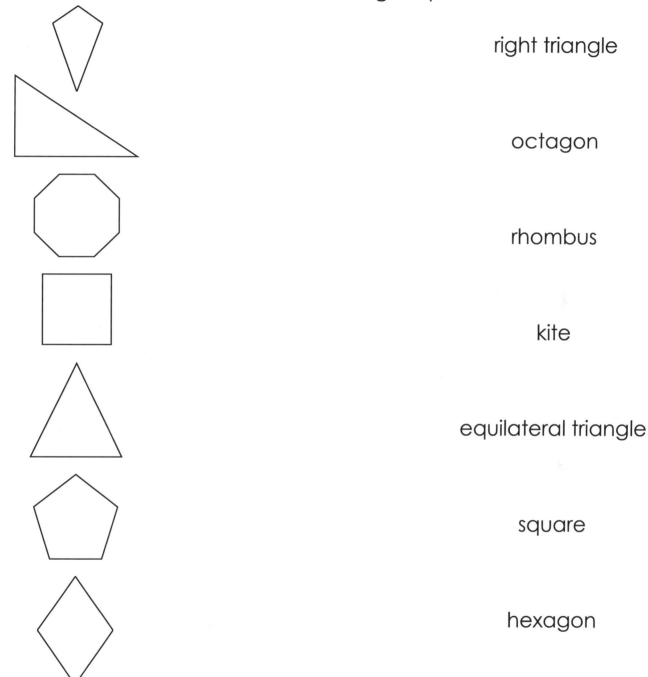

right triangle

octagon

rhombus

kite

equilateral triangle

square

hexagon

pentagon

trapezoid

The Lord helps _____ to study well.

3 Dimensional Solids - Spheres

A **circle** is round and flat.	A **sphere** is a round, full solid.
circle 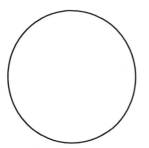	sphere 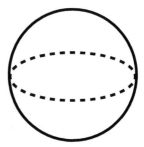

Each flat surface on a solid is a **face**. Each corner in a solid is a **vertex**. The sphere above has 0 faces and 0 vertices.

Look at the objects below. Color in all the sphere solids.

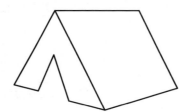

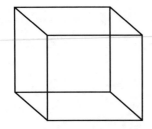

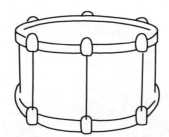

Circle or Sphere

Draw a line from each picture to show
whether it is a circle or a sphere.

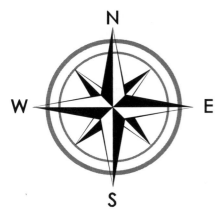

circle

sphere

_____ is a child of the Most High God.

3 Dimensional Solids - Cubes

A **square** is flat and has a length and width. square 	A **cube** is a solid and has a length, width, and height. cube 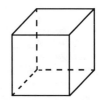

Each flat surface on a solid is a face. Each corner on a solid is a vertex. The cube above has 6 faces and 8 vertices.
Can you count them?

Write an **F** on each face of this cube. How many **F's** did you write?

_____ faces

Circle each vertex. How many vertices did you circle?

_____ vertices

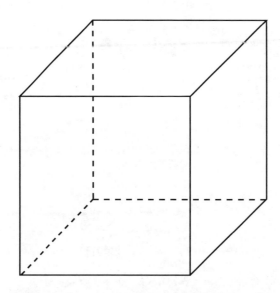

What shape does each face of a cube look like? _____

Identifying Cubes and Spheres

Color all the cubes in blue and the spheres in red.

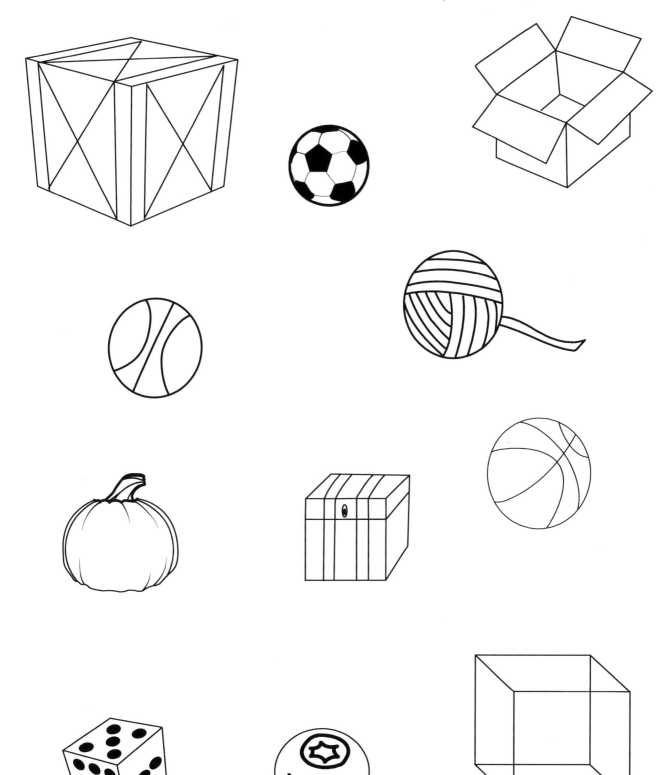

_____ is a child of the Most High God.

3 Dimensional Solids – Pyramids and Cones

A **pyramid** is a square base with triangle-shaped faces meeting at one top point. 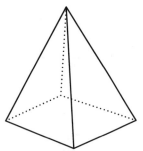	A **cone** is a circle base with a rounded piece meeting at one top point. 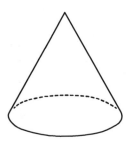
The pyramid above has 5 faces: 4 on the sides and 1 flat surface on the bottom. It has 5 vertices: 4 on the bottom and 1 at the top.	The cone above has 1 face at the bottom of the solid. It has 1 vertex at the top.

Draw a pyramid.	Draw a cone.

How many faces do you see?

_____

How many faces are there? ____

How many vertices do you see?

_____

How many vertices are there?

_____

How many faces do you see?

_____

How many faces are there? ____

How many vertices do you see?

_____

How many vertices are there?

_____

Identifying 3 Dimensional Solids

Look at each object and circle what type of solid it is.

	pyramid	sphere	cube
	cone	cube	pyramid
	sphere	cone	cube
	cube	sphere	pyramid
	cone	pyramid	cube
	pyramid	cube	sphere
	sphere	cone	pyramid

The Lord gives wisdom to_____.

3 Dimensional Solids - Cylinders and Triangular Prisms

A **cylinder** is a circle with the added dimension of height. 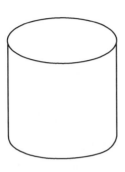	A **triangular prism** is a triangle with the added dimension of height. 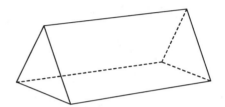
The cylinder above has 2 faces: 1 on the top and 1 on the bottom. It has no vertices or corners.	The prism above has 5 faces. It has 6 vertices. Can you count them?

Can cylinders be stacked on top of each other? _____

Can they roll? _____

Can you find a cylinder in your house? _____

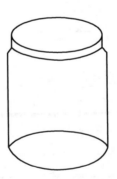

Can triangular prisms be stacked on top of each other? _____

Can they roll? _____

Can you find a triangular prism in your house? _____

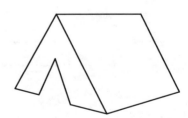

Identifying 3 Dimensional Solids

In each set, circle the correct solid.

Which solid has more faces?	Which solid has more vertices?

Which solid has more faces?	Which solid has more vertices?
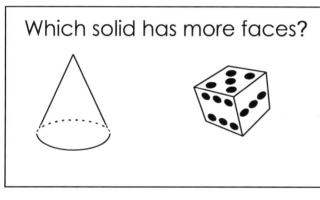	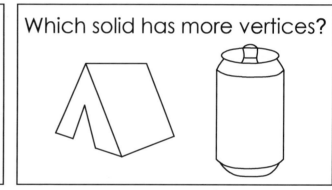

Which solid has more faces?	Which solid has more vertices?
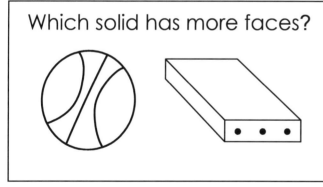	

Which solid has more faces?	Which solid has more vertices?
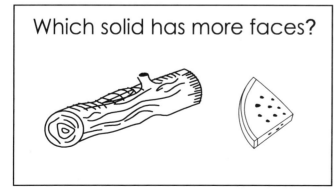	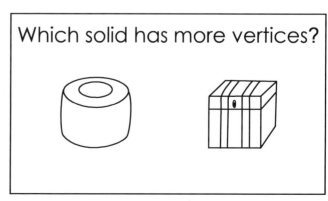

The Lord loves _____ with the greatest love.

3 Dimensional Solids - Review

Fill out the chart below for each 3 dimensional solid listed.

SOLID	NUMBER OF FACES	NUMBER OF VERTICES	DRAW IT
pyramid			
sphere			
cylinder			
cube			
triangular prism			
cone			

Identifying 3 Dimensional Solids

Color in all the solids as shown below.

cubes - blue   cylinders - yellow   triangular prisms - red

spheres - pink   cones - green   pyramids - purple

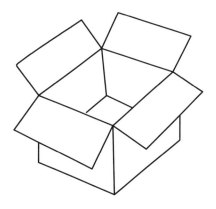
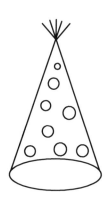

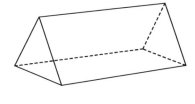
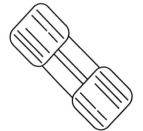
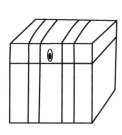
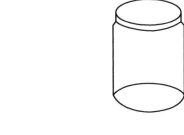
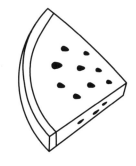

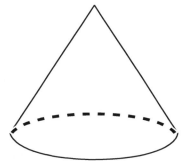

Vocabulary Cards

Unit 9 Vocabulary Cards

Division

6 apples divided into 2 equal groups=
3 apples in each group
6 ÷ 2 = 3
divide things into groups

Dividend

$$6 \div 2 = 3$$

→ Dividend

a number to be divided by another number

Divisor

$$6 \div 2 = 3$$

← Divisor

a number that is divided into
the dividend

Quotient

$$6 \div 2 = 3$$

Quotient →

result by dividing one quantity by another

Identity property

$$2 \div 1 = 2$$

a number ÷ 1 is the same number

Unit 10 Vocabulary Cards

Fractions

1 whole circle

2 of the 4 parts are shaded

$$\frac{2}{4}$$

part of an equal amount

Numerator

 $\longrightarrow \frac{2}{4}$

the top number in a fraction

Denominator

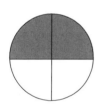 $\longrightarrow \frac{2}{4}$

the bottom number in a fraction

Mixed numbers

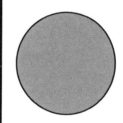 $1 \frac{1}{2}$

whole numbers AND parts of items

Half - 1/2

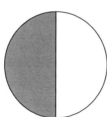

1 part of 2 are shaded

$$\frac{1}{2}$$

2 equal parts of a whole

Fourths / quarters - 1/4

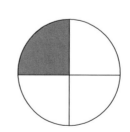

1 part of 4 are shaded

$$\frac{1}{4}$$

parts of a whole or group

Unit 10 Vocabulary Cards

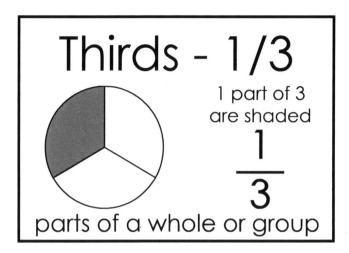

Thirds - 1/3

1 part of 3
are shaded

$$\frac{1}{3}$$

parts of a whole or group

Unit 11 Vocabulary Cards

Decimal point

Whole dollars | Parts of a dollar

$4.25

separates a whole number from the
fractional part of a number

tenths - Dimes

10   =   1
dimes   dollar
equal to 1 tenth of a dollar

hundreds - Pennies

100   =   1
pennies   dollar
equal to 1 hundredth of dollar

Place value of Decimals
tens, ones, tenths, hundredths

tens   tenths

7 3 . 5 2

ones ↑   hundredths

value is recognized by position of the dot

Fractions and Decimals
1 3/4 = 1.75

different way to write same amount

Unit 13 Vocabulary Cards

a.m.

Things I do in the a.m.
1. sleep
2. eat breakfast
3. watch sunrise
4. get ready for school
5. play outside

the period from midnight until noon

p.m.

Things I do in the p.m.
1. eat lunch
2. finish my school work
3. play outside
4. eat dinner and help clean up
5. take a bath and go to bed

the period from noon until midnight

quarter past

1/4 of an hour = 15 minutes

half past

1/2 of an hour = 30 minutes

Quarter to - :45

3/4 of an hour = 45 minutes

o'clock - :00

midnight or 12 noon

Unit 14 Vocabulary Cards

degrees

a unit of measurement used for temperature

Celsius

freezing point is 0°C

boiling point is 100°C

one of several temperature scales

Fahrenheit

freezing point is 32°F

boiling point is 212°F

one of several temperature scales

Standard system

inches

feet

yards

one group of units of measurement

inches

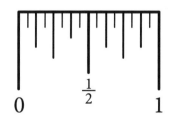

$$0 \qquad \frac{1}{2} \qquad 1$$

a unit(s) of length

feet

12 inches = 1 foot
24 inches = 2 feet

interesting fact: a dollar bill is 6 inches

a unit for measuring length

Unit 14 Vocabulary Cards

yard

12 inches = 1 foot
36 inches = 3 feet
3 feet = 1 yard

an English unit of length

Metric System

meter	1
decimeter	1/10
centimeter	1/100
millimeter	1/1000

base ten system of measurement

millimeter

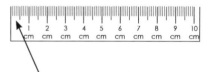

10 millimeters in
one centimeter

smallest unit of metric measurement

centimeter

A unit of length in the metric
system which is equal to 0.01
meter or 1/100 of 1 meter

unit of measurement equal to 10mm

meter

A yard stick is 36 inches and a
meter stick is 100 centimeters.

A yard stick is a little shorter
than a meter stick.

unit of measurement equal to 100cm

Unit 15 Vocabulary Cards

pounds

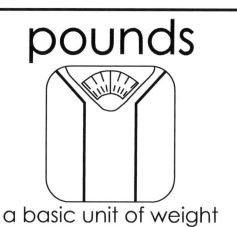

a basic unit of weight

ounces

a weight equal to 1/16 of a pound

Standard weight

Standard weight is measured in:

ounces
pounds
tons

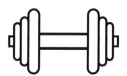

a system used to measure heaviness

tons

large items are measured in tons.

Some Elephants can weigh as much as 7 tons.

1 ton = 2000 pounds

Metric weight

Metric weight is measured in:

grams
kilograms

an additional system used to measure heaviness

gram

The weight of a regular paper clip is about 1 gram

a small metric unit used to measure

Unit 15 Vocabulary Cards

kilogram

A quart of milk has a weight of about 1 kilogram

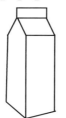

a larger metric unit used to measure

Standard capacity

liquids are measured in ounces, cups, pints, quarts, and gallons

quantity a container can hold when full

fluid ounces

Standard can contains 12 fluid ounces

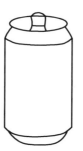

a small unit of liquid measurement

cup

1 cup = 8 fluid ounces

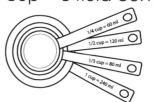

1/4 cup = 60 ml
1/2 cup = 120 ml
1/3 cup = 80 ml
1 cup = 240 ml

a unit of measurement

pint

1 pint = 16 fluid ounce
1 pint = 2 cups or about half a liter

a unit of measurement

quart

A quart is a quarter of a gallon!
1 quart = 2 pints
1 quart = 4 cups
1 quart = 32 fluid ounces

a unit of measurement

Unit 15 Vocabulary Cards

gallon

1 gallon = 4 quarts
1 gallon = 8 pints
1 gallon = 16 cups
1 gallon = 128 fluid
ounces

a unit of measurement

Metric capacity

liquids are
measured in
milliliter and liter.

total amount of liquid an object can contain

milliliter

A milliliter is a very small amount
of liquid. It only fills the bottom
of the teaspoon.

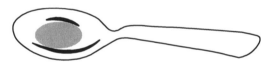

A metric unit used to measure capacity

liter

A liter can
be found
in different
shapes.

a metric unit of measurement

Unit 16 Vocabulary Cards

line graph

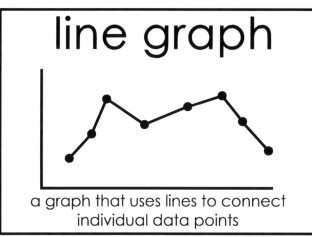

a graph that uses lines to connect individual data points

pie chart

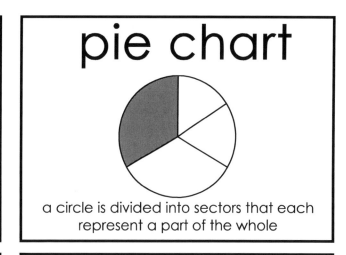

a circle is divided into sectors that each represent a part of the whole

bar chart

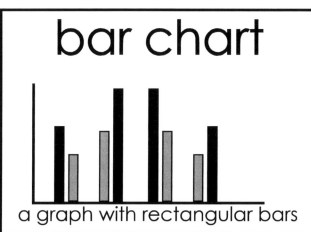

a graph with rectangular bars

pictograph

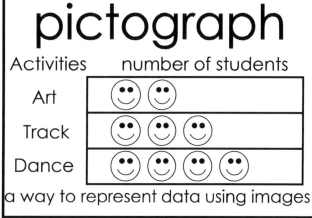

a way to represent data using images

title of a map

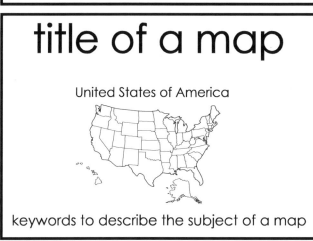

keywords to describe the subject of a map

key of a map

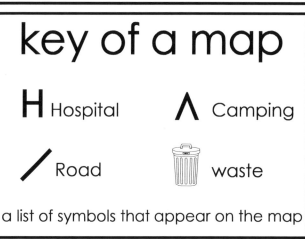

a list of symbols that appear on the map

Unit 16 Vocabulary Cards

scale

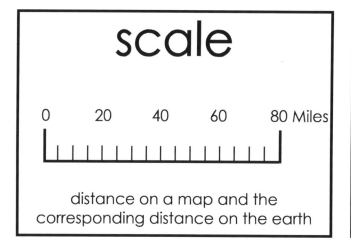

distance on a map and the corresponding distance on the earth

compass rose

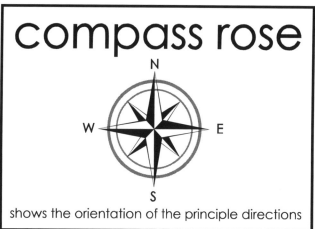

shows the orientation of the principle directions

Unit 17 Vocabulary Cards

quadrilateral

a shape which has four straight sides

parallelogram

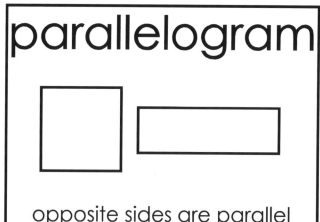

opposite sides are parallel

square

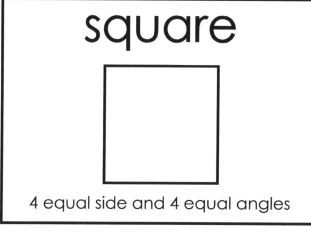

4 equal side and 4 equal angles

rectangle

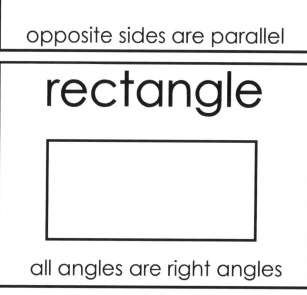

all angles are right angles

pentagon

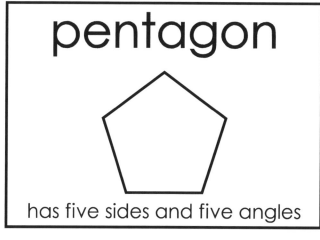

has five sides and five angles

trapezoid

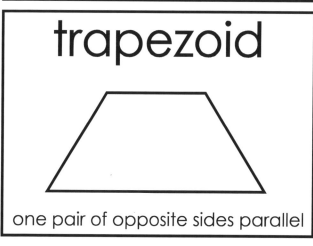

one pair of opposite sides parallel

Unit 17 Vocabulary Cards

perimeter

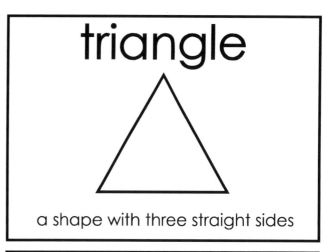

the total sum of all sides

triangle

a shape with three straight sides

symmetry

A shape has symmetry if a central dividing line can be drawn on it, to show that both sides of the shape are exactly the same.

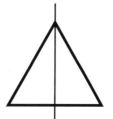

the same on both sides

hexagon

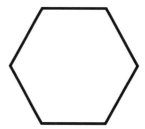

a shape with six straight sides

octagon

made up of 8 sides

rhombus

four equal sides like a diamond

Unit 17 Vocabulary Cards

## equilateral triangle 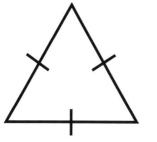 all sides are equal in length	## isosceles triangle  two sides are the same
## scalene triangle 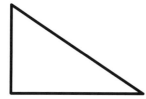 has three unequal sides	## right triangle A triangle in which one of the interior angles is 90° one angle is 90 degrees
## Kite  has two pairs of sides = in length	## sphere 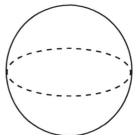 3-dimensional object shaped like a ball

Unit 17 Vocabulary Cards

cylinder

A cylinder is a three-dimensional solid figure which consists of two circular bases connected with a curved surface made by folding a rectangle.

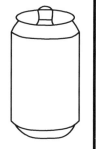

like a soda can

cube

Objects that are cube-shaped include building blocks and dice.

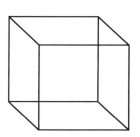

contained by six equal squares.

cone

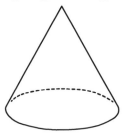

continuous curved surface tapering to a point.

pyramid

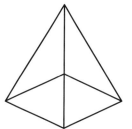

sides are triangles which meet at the top

triangular prism

3-D solid with two triangle bases and three rectangular sides.

face

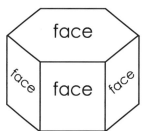

Any of the individual flat surfaces of a solid object.

Unit 17 Vocabulary Cards

vertex or vertices

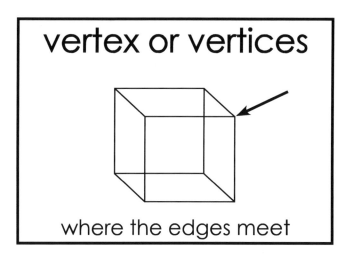

where the edges meet

Answer Keys

Unit 9 Week 19 Answer Key

Introduction to Division

Division Practice

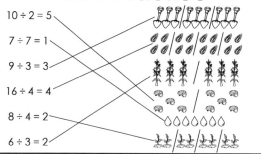

$10 \div 2 = 5$

$7 \div 7 = 1$

$9 \div 3 = 3$

$16 \div 4 = 4$

$8 \div 4 = 2$

$6 \div 3 = 2$

Dividing into Groups

_____8_____ in each group

_____2_____ in each group

Dividing into Groups

$12 \div 4 =$ __3__

$20 \div 2 =$ __10__

$24 \div 6 =$ __4__

$30 \div 5 =$ __6__

$15 \div 3 =$ __5__

Divide and Multiply

$4 \times 1 = 4$

$4 \div 4 = 1$

$4 \times 3 = 12$

$12 \div 4 = 3$

$2 \times 4 = 8$

$8 \div 2 = 4$

$5 \times 2 = 10$

$10 \div 5 = 2$

Division Mix-Up

$24 \div 4 = 6$

$18 \div 2 = 9$

$16 \div 4 = 4$

$21 \div 3 = 7$

$12 \div 4 = 3$

$30 \div 10 = 3$

$8 \div 4 = 2$

$15 \div 5 = 3$

$14 \div 7 = 2$

$30 \div 6 = 5$

$7 \div 7 = 1$

$20 \div 4 = 5$

Division Vocabulary

$12 \div 3 = 4$

$\dfrac{12}{\boxed{3}} = 4$

$\boxed{15} \div 5 = 3$

$8 \div \boxed{2} = 4$

$10 \div 2 = \boxed{5}$

$\dfrac{10}{2} = 5$

$\dfrac{15}{5} = \boxed{3}$

$\dfrac{\boxed{8}}{2} = 4$

Repeated Subtraction

$15 \div 5$

$\begin{array}{r} 15 \\ -5 \\ \hline \end{array} = 10$ $\begin{array}{r} 10 \\ -5 \\ \hline \end{array} = 5$ $\begin{array}{r} 5 \\ -5 \\ \hline \end{array} = 0$

How many times did you subtract? __3__ $15 \div 5 =$ __3__

$20 \div 4$

$\begin{array}{r} 20 \\ -4 \\ \hline \end{array} = 16$ $\begin{array}{r} 16 \\ -4 \\ \hline \end{array} = 12$ $\begin{array}{r} 12 \\ -4 \\ \hline \end{array} = 8$ $\begin{array}{r} 8 \\ -4 \\ \hline \end{array} = 4$ $\begin{array}{r} 4 \\ -4 \\ \hline \end{array} = 0$

How many times did you subtract? __5__ $20 \div 4 =$ __5__

$16 \div 8$

$\begin{array}{r} 16 \\ -8 \\ \hline \end{array} = 8$ $\begin{array}{r} 8 \\ -8 \\ \hline \end{array} = 0$

How many times did you subtract? __2__ $16 \div 8 =$ __2__

$24 \div 6$

$\begin{array}{r} 24 \\ -6 \\ \hline \end{array} = 18$ $\begin{array}{r} 18 \\ -6 \\ \hline \end{array} = 12$ $\begin{array}{r} 12 \\ -6 \\ \hline \end{array} = 6$ $\begin{array}{r} 6 \\ -6 \\ \hline \end{array} = 0$

How many times did you subtract? __4__ $24 \div 6 =$ __4__

Division by 2

$10 \div 2 = 5$ $16 \div 2 = 8$ $20 \div 2 = 10$

$8 \div 2 = 4$ $12 \div 2 = 6$ $14 \div 2 = 7$ $4 \div 2 = 2$

$18 \div 2 = 9$ $2 \div 2 = 1$ $22 \div 2 = 11$ $6 \div 2 = 3$

Color by Number - Division

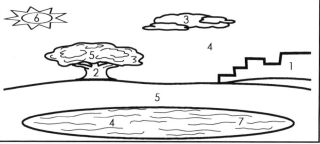

Unit 9 Week 20 Answer Key

Division Recap

Enjoy dividing items around the house into groups.

Division Recap

Division is repeated subtraction.

Division is splitting things into equal sized groups.

Quotients are the answers to division problems.

Dividends are the total numbers that are divided into groups.

Dividing by 2 is dividing a group in half.

(24) ÷ 6 = 4

35 ÷ 7 = (5)

30 ÷ 6
30	24	18	12	6
-6	-6	-6	-6	-6
24	18	12	6	0

18 ÷ (2) = 9

16 can be split into 4 groups of 4.

Division Math Stories

When the girls cleaned their room, they <u>divided</u> the 16 shirts into 4 <u>groups</u>. How many are in <u>each</u> group?

Wendi and Gary put the 20 books into 5 <u>piles</u>. How many books are in <u>each</u> pile?

The 12 moving boxes were filled and put on the truck in 3 <u>rows</u>. How many boxes are in <u>each</u> row?

Mom made 8 different bowls of modeling clay. The 2 brothers shared it all <u>equally</u>. How many did <u>each</u> one get?

More Division Stories

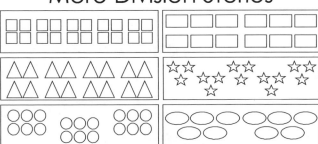

Dividing Into 3 Groups

How many dogs are in each bed? _____3_____

We write this as: 9 ÷ 3 = __3__

Dividing by 3

How many fruit snacks does each friend have? _____10_____

We write this as: 30 ÷ 3 = __10__

How many crayons does each table have? __20__

We write this as: 60 ÷ 3 = __20__

Division Word Stories

12 ÷ 2 = 6 $\frac{12}{2}$ = 6

10 ÷ 2 = 5 $\frac{10}{2}$ = 5

20 ÷ 2 = 10 $\frac{20}{2}$ = 10

14 ÷ 2 = 7 $\frac{14}{2}$ = 7

Division Word Stories

(24) ÷ 3 = 8 $\frac{(24)}{3}$ = 8   (10) ÷ 5 = 2 $\frac{(10)}{5}$ = 2

(21) ÷ 3 = 7 $\frac{(21)}{3}$ = 7   (25) ÷ 5 = 5 $\frac{(25)}{5}$ = 5

(15) ÷ 3 = 5 $\frac{(15)}{3}$ = 5   (5) ÷ 5 = 1 $\frac{(5)}{5}$ = 1

(9) ÷ 3 = 3 $\frac{(9)}{3}$ = 3   (15) ÷ 5 = 3 $\frac{(15)}{5}$ = 3

Multiplying and Dividing by 1

$\frac{5}{1}$ = 5 $\frac{6}{1}$ = 6   2 ÷ 1 = 2   7 x 1 = 7

12 ÷ 1 = 12   10 x 1 = 10 $\frac{8}{1}$ = 8   3 x 1 = 3

4 x 1 = 4   9 ÷ 1 = 9   14 ÷ 1 = 14 $\frac{11}{1}$ = 11

Dividing into Groups of 3, 4, 5

How many plates will she give away? __6__   30 ÷ 5 = __6__

How many shelves will he use for his cars? __5__   20 ÷ 4 = __5__

How many cans will she use for the tennis balls? __7__   21 ÷ 3 = __7__

Unit 10 Week 21 Answer Key

Fraction Basics

$$\frac{2}{8}$$ $$\frac{5}{8}$$

$$\frac{6}{8}$$ $$\frac{3}{8}$$

$$\frac{1}{8}$$ $$\frac{4}{8}$$

Fraction Basics

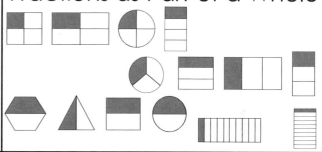

Fractions as Part of a Set

Fractions as Part of a Whole

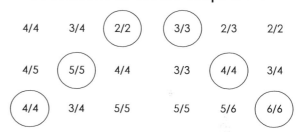

Fractions that Equal 1

$$\frac{\boxed{1}}{\boxed{1}} = 1$$ $$1 = \frac{\boxed{3}}{\boxed{3}}$$

$$\frac{\boxed{4}}{\boxed{4}} = 1$$ $$1 = \frac{\boxed{5}}{\boxed{5}}$$

Fractions that Equal 1

4/4   3/4   (2/2)   (3/3)   2/3   2/2

4/5   (5/5)   4/4   3/3   (4/4)   3/4

(4/4)   3/4   5/5   5/5   5/6   (6/6)

Mixed Numbers

Comparing Fraction Amounts

$$\left(\frac{4}{6}\right) \quad \frac{1}{3}$$ $$\left(\frac{3}{4}\right) \quad \frac{1}{2}$$

$$\left(\frac{3}{5}\right) \quad \frac{2}{4}$$ $$\frac{1}{2} \quad \left(\frac{2}{3}\right)$$

Fraction Review

There are ___3___ total faces. ___2___ of them are shaded. $$\frac{\boxed{2}}{\boxed{3}}$$

There are ___5___ total wrenches. ___4___ of them are shaded. $$\frac{\boxed{4}}{\boxed{5}}$$

There are ___2___ total worms. ___1___ of them are circled. $$\frac{\boxed{1}}{\boxed{2}}$$

There are ___4___ total burgers. ___1___ of them are circled. $$\frac{\boxed{1}}{\boxed{4}}$$

There are ___5___ total ladders ___2___ of them are circled. $$\frac{\boxed{2}}{\boxed{5}}$$

There are ___6___ total drums. ___3___ of them are circled. $$\frac{\boxed{3}}{\boxed{6}}$$

Fraction Review

The numerator is the top number of a fraction.

A fraction with matching numbers is equal to 1.

A fraction can be written with a straight line or a slanted one.

A mixed number shows some whole amounts and some fraction amounts.

The denominator is the bottom number of a fraction.

$$3\frac{1}{4}$$

$$\left(\frac{5}{6}\right)$$

$$\frac{3}{3} = 1$$

$$\frac{2}{\left(3\right)}$$

$$3/4 = \frac{3}{4}$$

Unit 10 Week 22 Answer Key

Fraction Word Stories

one quarter ⟶ 3/4
two thirds ⟶ 1/4
three fourths ⟶ 2/2
two halves ⟶ 2/3

Half, Thirds, and Fourths

$\frac{1}{3}$

$\frac{1}{2}$

$\frac{1}{4}$

$\frac{2}{3}$

$\frac{3}{4}$

$\frac{1}{2}$

$\frac{2}{4}$

Fraction Word Stories

$\frac{1}{12}$

$\frac{2}{3}$

$\frac{5}{8}$

$\frac{8}{11}$

$\frac{1}{10}$

More Fraction Stories

$\frac{7}{8}$

$\frac{11}{15}$

$\frac{4}{5}$

$\frac{10}{12}$

$\frac{32}{}$

$\frac{11}{32}$

Fractional Words

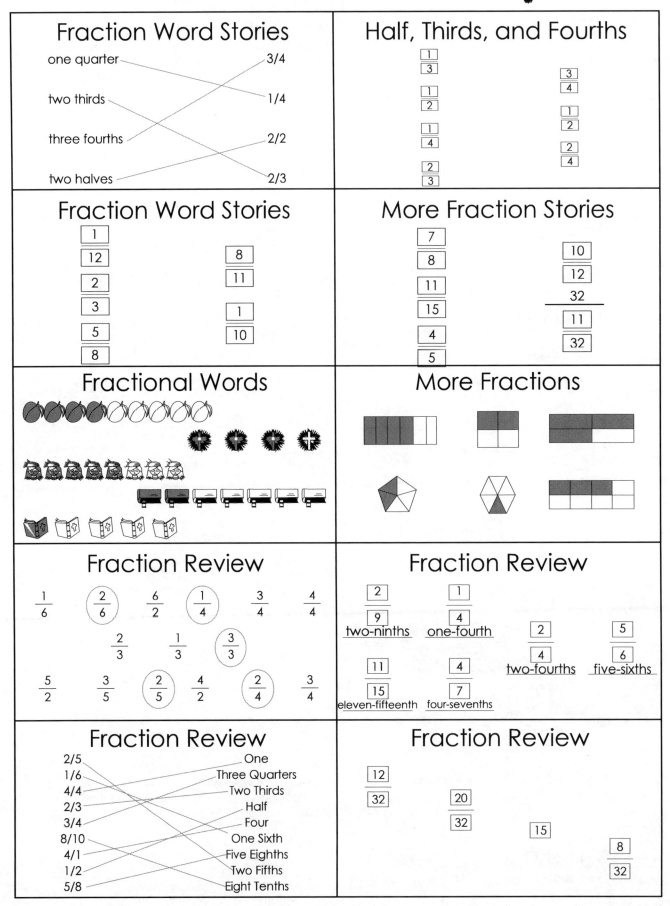

More Fractions

Fraction Review

$\frac{1}{6}$ ⬭$\frac{2}{6}$⬭ $\frac{6}{2}$ ⬭$\frac{1}{4}$⬭ $\frac{3}{4}$ $\frac{4}{4}$

$\frac{2}{3}$ $\frac{1}{3}$ ⬭$\frac{3}{3}$⬭

$\frac{5}{2}$ $\frac{3}{5}$ ⬭$\frac{2}{5}$⬭ $\frac{4}{2}$ ⬭$\frac{2}{4}$⬭ $\frac{3}{4}$

Fraction Review

$\frac{2}{9}$ two-ninths

$\frac{1}{4}$ one-fourth

$\frac{2}{4}$ two-fourths

$\frac{5}{6}$ five-sixths

$\frac{11}{15}$ eleven-fifteenth

$\frac{4}{7}$ four-sevenths

Fraction Review

2/5 ⟶ One
1/6 ⟶ Three Quarters
4/4 ⟶ Two Thirds
2/3 ⟶ Half
3/4 ⟶ Four
8/10 ⟶ One Sixth
4/1 ⟶ Five Eighths
1/2 ⟶ Two Fifths
5/8 ⟶ Eight Tenths

Fraction Review

$\frac{12}{32}$

$\frac{20}{32}$

15

$\frac{8}{32}$

Unit 11 Week 23 Answer Key

Introduction to Decimals

$5.71 $9.18

$1.32 $2.99

$3.45 $8.27

$6.83 $7.50

Decimal Place Values

Dollars	Tenths	Hundredths
9	1	5
3	9	1
1	8	9
8	7	7
2	2	9
5	4	6
7	3	4

Decimals - Whole and Parts

Decimals - Whole and Parts

Comparing Decimals

25.62	>	25.10	11.55	>	10.92
67.12	>	65.14	33.21	<	33.83
79.42	>	70.45	84.67	=	84.67
42.26	>	62.42	93.52	>	92.90
52.51	=	52.51	0.19	<	1.00
3.81	>	3.18	17.32	=	71.32

Number Line Decimals

Decimal and Fraction Amounts

Fraction	Decimal
3/4	.75
2 1/4	2.25
2 1/2	2.50
1 1/4	1.25

Decimal and Fraction Amounts

3.50	1 3/4
0.75	1 1/4
3.25	3 1/2
1.75	1/2
3.75	3/4
2.50	3 1/4
0.50	3 3/4
1.25	2 1/2

Decimals Review

63.45	tens	(ones)	tenths	hundredths
10.29	(tens)	ones	tenths	hundredths
37.51	tens	ones	tenths	(hundredths)
14.57	tens	ones	(tenths)	hundredths
28.89	tens	ones	tenths	(hundredths)
45.96	tens	ones	tenths	(hundredths)

Decimals Review

34.23	31.90	50.05	50.05	14.00	14.00
32.86	32.86	51.90	50.75	19.60	14.11
39.12	34.23	50.75	50.98	17.45	17.45
31.90	39.12	50.98	51.90	14.11	19.60
91.36	90.90	12.09	12.00	65.16	60.56
92.60	91.36	21.90	12.09	60.56	62.10
90.90	91.63	12.00	20.92	62.10	65.16
91.63	92.60	20.92	21.90	65.65	65.65

Unit 11 Week 24 Answer Key

Decimals

5.67 > 5. <u>66 or below</u>   6.17 < 6. <u>18 or above</u>

1.90 < 1. <u>91 or above</u>   3.12 > 3. <u>11 or below</u>

8.09 = 8. <u>09</u>   2.24 < 2. <u>25 or above</u>

4.87 < 4. <u>88 or above</u>   9.25 > 9. <u>24 or below</u>

Whole and Part Decimals

1.75

0.25

2.50

3.25

1.50

Decimal Tenths

<u>Have fun counting dimes.</u>

Tenths

This is __6__ tenths.

This is __1__ tenths.

This is __8__ tenths.

This is __5__ tenths.

This is __3__ tenths.

Decimal Hundredths

	Whole Dollars	Dimes - Tenths	Pennies - Hundredths
1.94	1	9	4
2.27	2	2	7
5.03	5	0	3
0.90		9	0

Rounding to Tenths

5.29	5.3	6.89	6.9
2.13	2.1	9.74	9.7
3.91	3.9	7.55	7.6
1.48	1.5	4.62	4.6

Adding Decimals

49.49   36.94   39.94

76.54   66.97   73.69

89.57   38.89   36.79

77.73   75.79   51.95

Adding Tenths and Hundredths

6.97
__5__ hundredths + __2__ hundredths = __7__ hundredths
__1__ tenths + __8__ tenths = __9__ tenths
__4__ whole + __2__ whole = __6__ whole

4.78
__1__ hundredths + __7__ hundredths = __8__ hundredths
__7__ tenths + __0__ tenths = __7__ tenths
__1__ whole + __3__ whole = __4__ whole

7.79
__9__ hundredths + __0__ hundredths = __9__ hundredths
__6__ tenths + __1__ tenths = __7__ tenths
__5__ whole + __2__ whole = __7__ whole

10.77
__4__ hundredths + __3__ hundredths = __7__ hundredths
__2__ tenths + __5__ tenths = __7__ tenths
__2__ whole + __8__ whole = __10__ whole

Decimal Review

9.5   <   tenths

>   1.2   4.15

hundredths   0.75   3.8

Adding Decimal Review

2.05	2.10	3.2
.5	1.02	1.2
2.05	1.4	1.04
1.5	1.04	1.40
1.25	1.50	3.5
.75	0.50	0.3

Unit 12 Week 25 Answer Key

Adding Decimals in Money

$378.79	$986.39
$79.98	$98.79
$724.77	$557.76
$179.92	$277.25

Addition with Decimals

$42.50
+ $31.32
= $73.82
 $ 3.49
+ $90.17
= $93.66

$29.64
+ $ 5.24
= $34.88
 $ 68.03
+ $171.86
= $239.89

Picnic Math Stories

$ 7.50
+ $13.20
= $20.70
 $ 8.75
+ $ 3.80
= $12.55

$ 5.36
+ $ 3.21
= $ 8.57

$11.47
+ $ 4.59
= $16.06
 $ 9.18
+ $ 2.95
= $ 12.13

Decimal Addition

$15.60 Rounds to:
+$ 8.25
$23.85 $24.00

$ 5.75 Rounds to:
+$ 3.50
$ 9.25 $9.00

$ 5.90 Rounds to:
+$12.85
$18.75 $19.00

$ 8.00 Rounds to:
+$ 4.20
$12.20 $12.00

$ 1.97 Rounds to:
+$ 2.44
$ 4.41 $ 4.00

Decimals in Money Stories

$ 4.25
+$ 11.87
= $ 16.12

$ 6.15
+$10.52
= $16.67

$ 5.72
+$15.29
= $21.01

Whole dollars	16	Whole dollars	16
Quarters		Quarters	2
Dimes	1	Dimes	1
Nickels		Nickels	1
Pennies	2	Pennies	2

Whole dollars 21
Quarters
Dimes
Nickels
Pennies 1

Addition with Decimals

$ 315.73
+ $ 152.17

$ 71.48
+ $ 215.31

$ 127.56
+ $ 171.22

$ 337.04
+ $ 51.93

$ 160.10
+ $ 208.06

$ 298.78
$ 467.90
$ 368.16
$ 286.79
$ 388.97

Decimal Addition

$ .80
+$ .39
= $1.19

$1.00
+$ .55
= $ 1.55

$ .60
+$1.04
= $ 1.64

Decimal Addition Stories

4.6
+ 11.2
= 15.8

2.25
+ 3.50
= 5.75

6.30
+ 9.75
= 16.05

8.8
+ 7.4
= 16.2

5.75   15.8   16.05   16.2
least   greatest

Addition with Money Stories

Sum = $12.02
11 dollars, 2 pennies   (12 dollars, 2 pennies)   12 dollars, 2 dimes, 1 penny

Sum = $5.90
(5 dollars, 3 quarters, 1 dime, 1 nickel)   5 dollars, 3 quarters, 2 dimes   5 dollars, 3 quarters, 1 nickel

Sum = $13.10
13 dollars, 1 nickel   12 dollars, 1 dime   (13 dollars, 1 dime)

Sum = $10.13
(10 dollars, 1 dime, 3 pennies)   10 dollars, 1 nickel, 3 pennies   9 dollars, 3 quarters, 2 dimes, 3 pennies

Decimal Addition Review

$5.50
+$ .75
= $6.25

$9.00
+$ .24
= $9.24

$7.21
+$ .36
= $7.57

$4.34
+$1.30
= $5.64

$8.13
+$1.34
= $9.47

$14.10
+$ .30
= $14.40

Unit 12 Week 26 Answer Key

Subtraction with Decimals

7.50 - 5.23 2.27	1.70 - 0.94 0.76	5.60 - 3.38 2.22
2.90 - 1.36 1.54	5.40 - 4.15 1.25	9.80 - 7.67 2.13
3.30 - 0.29 3.01	8.90 - 6.71 2.19	6.50 - 3.42 3.08

Subtracting Decimals

4.5 -3.2 1.3	4.54 - 3.2 4.22	4.54 -3.2 1.34	4.54 -3.20 1.34	9.70 -3.4 6.30	9.7 -3.4 6.3	9.70 - 3.4 9.36	9.70 -3.40 6.30
6.05 - 4.3 5.62	6.05 -4.3 1.75	6.05 -4.30 1.75	6.05 -4.30 1.75	5.8 -2.6 3.2	5.80 -2.6 3.20	5.80 - 2.6 5.54	5.80 -2.60 3.20
7.92 -5.7 2.22	7.92 - 5.7 7.35	7.92 5.70 2.22	7.92 -5.70 2.22	8.6 -1.5 7.1	8.60 - 1.5 8.45	8.60 -1.5 7.10	8.60 -1.50 7.10

Decimal Subtraction Stories

What is the <u>difference</u> between the lengths?
1.8

<u>How many more</u> feet do the regular swings reach?
5.2

<u>How much</u> of the fence was left standing?
248

<u>How many more</u> degrees was it yesterday than today?
4.6

Subtraction Stories

6.75 -2.50 4.25	10.0 - 2.2 7.8	9.4 - 4.5 4.9
3.00 -1.75 1.25	6.8 -3.5 3.3	8.50 -7.75 0.75

Subtraction with Decimals

$96 = $96.00	$5 = $5.00	$45 = $45.00	$15 = $15.00
$70 = $70.00	$11 = $11.00	$20 = $20.00	$2 = $2.00
$10 = $10.00	$30 = $30.00	$60 = $60.00	$19 = $19.00
$50 = $50.00	$25 = $25.00	$83 = $83.00	$0 = $0.00

Subtracting Money

$4.80 - $2.35 $2.45	$520.65 - $200.30 $320.35
$1554.27 -$311.11 $1237.16	$ 68.03 - $171.86 = $239.89

Decimals are Parts of a Dollar

3.75 — 3 1/2
3.25 — 3 1/10
3.5 — 3 3/4
3.00 — 3 0/4
3.10 — 3 1/4

Subtraction Stories

45.00
-25.25
19.75

16.00
-13.75
2.25

22.00
-10.50
11.25

Decimal Subtraction Review

$7.11

$8.09

$7.91

2.25

1.5

2.5

3.75 $53.52

2.75 $53.48

3.25 $52.48

Decimal Review

swings. 0.6 + 0.6 + 0.6 + 0.6 + 0.6 + 0.6 + 0.6 = 4.20 hours

swings: 1.80 hours
slide: 1.25 hours
difference: 0.55 hours

the swings 0.6 + 0.6 + 0.6 + 0.6 = 2.4 hours

Saturday swings: 2.4 hours
Sunday swings: 1.8 hours
difference: 0.6 hours

Unit 13 Week 27 Answer Key

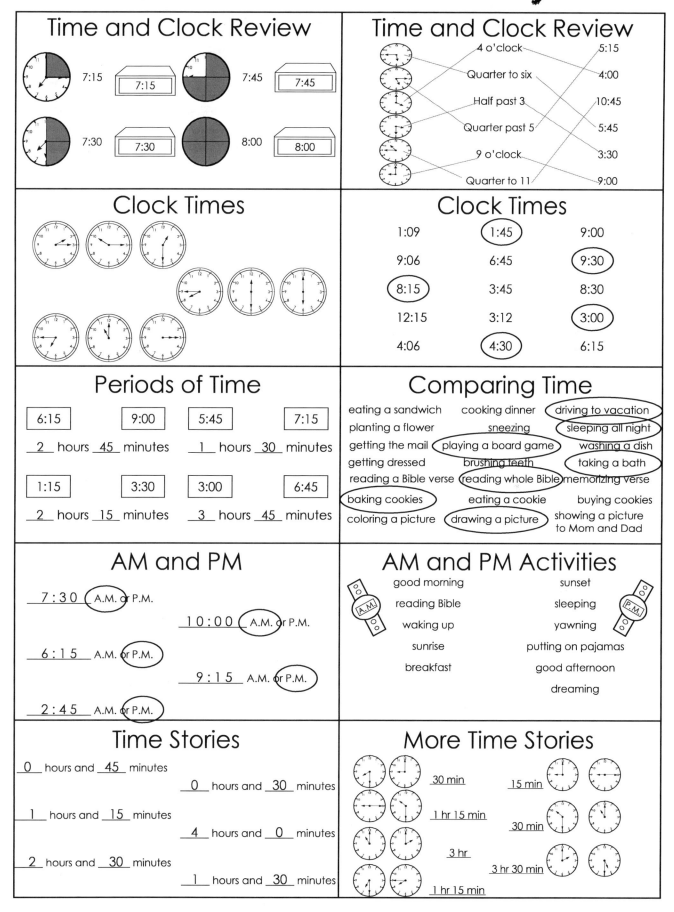

Time and Clock Review

7:15 | 7:15 |
7:45 | 7:45 |
7:30 | 7:30 |
8:00 | 8:00 |

Time and Clock Review

4 o'clock   5:15
Quarter to six   4:00
Half past 3   10:45
Quarter past 5   5:45
9 o'clock   3:30
Quarter to 11   9:00

Clock Times

Clock Times

1:09   (1:45)   9:00
9:06   6:45   (9:30)
(8:15) 3:45   8:30
12:15   3:12   (3:00)
4:06   (4:30)   6:15

Periods of Time

| 6:15 | 9:00 | 5:45 | 7:15 |

__2__ hours __45__ minutes   __1__ hours __30__ minutes

| 1:15 | 3:30 | 3:00 | 6:45 |

__2__ hours __15__ minutes   __3__ hours __45__ minutes

Comparing Time

eating a sandwich   cooking dinner   (driving to vacation)
planting a flower   sneezing   (sleeping all night)
getting the mail   (playing a board game)   washing a dish
getting dressed   brushing teeth   (taking a bath)
reading a Bible verse (reading whole Bible) memorizing verse
(baking cookies)   eating a cookie   buying cookies
coloring a picture (drawing a picture) showing a picture to Mom and Dad

AM and PM

__7:30__ (A.M.) or P.M.

__10:00__ (A.M.) or P.M.

__6:15__ A.M. (or P.M.)

__9:15__ A.M. (or P.M.)

__2:45__ A.M. (or P.M.)

AM and PM Activities

good morning   sunset
reading Bible   sleeping
waking up   yawning
sunrise   putting on pajamas
breakfast   good afternoon
                    dreaming

Time Stories

__0__ hours and __45__ minutes

__0__ hours and __30__ minutes

__1__ hours and __15__ minutes

__4__ hours and __0__ minutes

__2__ hours and __30__ minutes

__1__ hours and __30__ minutes

More Time Stories

__30 min__   __15 min__
__1 hr 15 min__   __30 min__
__3 hr__   __3 hr 30 min__
__1 hr 15 min__

Unit 13 Week 28 Answer Key

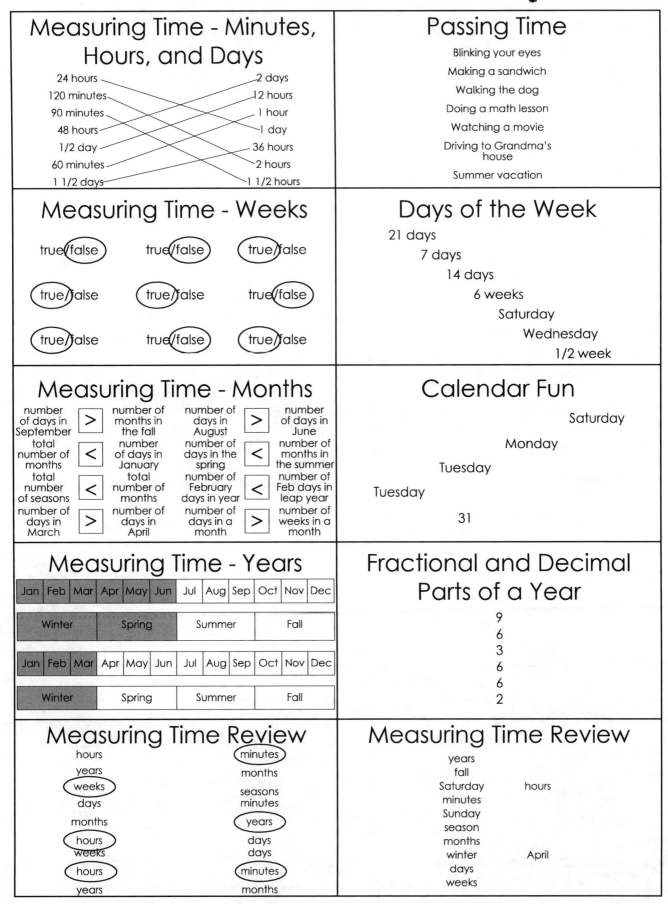

Measuring Time - Minutes, Hours, and Days

24 hours   2 days
120 minutes   12 hours
90 minutes   1 hour
48 hours   1 day
1/2 day   36 hours
60 minutes   2 hours
1 1/2 days   1 1/2 hours

Passing Time

Blinking your eyes
Making a sandwich
Walking the dog
Doing a math lesson
Watching a movie
Driving to Grandma's house
Summer vacation

Measuring Time - Weeks

true/**false**　　　**true**/false　　　**true**/false

true/false　　　true/**false**　　　true/**false**

true/**false**　　　true/**false**　　　true/**false**

Days of the Week

21 days
7 days
14 days
6 weeks
Saturday
Wednesday
1/2 week

Measuring Time - Months

number of days in September	>	number of months in the fall		number of days in August	>	number of days in June
total number of months	<	number of days in January		number of days in the spring	<	number of months in the summer
total number of seasons	<	total number of months		number of February days in year	<	number of Feb days in leap year
number of days in March	>	number of days in April		number of days in a month	>	number of weeks in a month

Calendar Fun

Saturday
Monday
Tuesday
Tuesday
31

Measuring Time - Years

| Jan | Feb | Mar | Apr | May | Jun | Jul | Aug | Sep | Oct | Nov | Dec |

| Winter | Spring | Summer | Fall |

| Jan | Feb | Mar | Apr | May | Jun | Jul | Aug | Sep | Oct | Nov | Dec |

| Winter | Spring | Summer | Fall |

Fractional and Decimal Parts of a Year

9
6
3
6
6
2

Measuring Time Review

hours   **minutes**
years   months
weeks   seasons
days   minutes
months   **years**
hours   days
weeks   days
hours   **minutes**
years   months

Measuring Time Review

years
fall
Saturday   hours
minutes
Sunday
season
months
winter   April
days
weeks

Unit 14 Week 29 Answer Key

Measuring Temperature

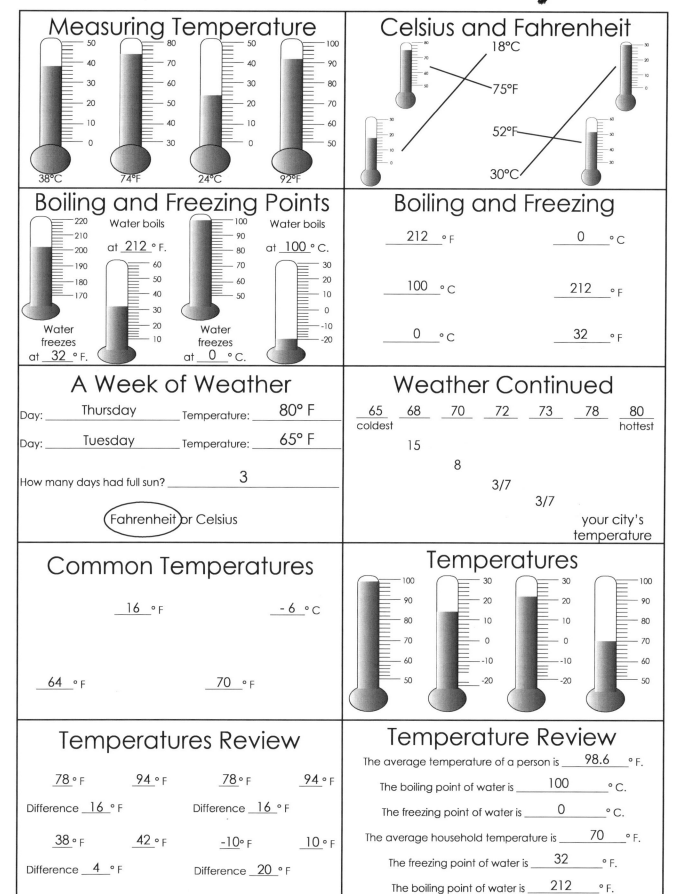

38°C 74°F 24°C 92°F

Celsius and Fahrenheit

18°C
75°F
52°F
30°C

Boiling and Freezing Points

Water boils at 212 ° F.

Water boils at 100 ° C.

Water freezes at 32 ° F.

Water freezes at 0 ° C.

Boiling and Freezing

212 ° F   0 ° C

100 ° C   212 ° F

0 ° C   32 ° F

A Week of Weather

Day: Thursday   Temperature: 80° F

Day: Tuesday   Temperature: 65° F

How many days had full sun? 3

Fahrenheit or Celsius

Weather Continued

65   68   70   72   73   78   80
coldest   hottest

15

8

3/7

3/7

your city's temperature

Common Temperatures

16 °F   - 6 °C

64 °F   70 °F

Temperatures

Temperatures Review

78 °F   94 °F   78 °F   94 °F

Difference 16 °F   Difference 16 °F

38 °F   42 °F   -10° F   10 °F

Difference 4 °F   Difference 20 °F

Temperature Review

The average temperature of a person is 98.6 °F.

The boiling point of water is 100 °C.

The freezing point of water is 0 °C.

The average household temperature is 70 °F.

The freezing point of water is 32 °F.

The boiling point of water is 212 °F.

Unit 14 Week 30 Answer Key

Measuring Length

Have fun learning about length.

How long is it?

6	inches	15	centimeters
3 3/4	inches	9.5	centimeters
4 3/4	inches	12	centimeters
1 1/2	inches	3.8	centimeters
3	inches	7.6	centimeters

Metric System

1 mm.   1 cm.

1 dm.   1 m.

Metric Measurements

5.5 cm.

7 cm.

3.5 cm.

9.5 cm.

2 cm.

8 cm.

Estimating Length

inches/centimeters

feet

yards/meters

How long is it?

(8 inches)   6 yards   (9 feet)   20 yards

(10 centimeters)   10 meters

12 inches   (20 feet)

10 inches   (100 yards)

(15 meters)   15 millimeters

Equal Measurements

24 inches = __2__ feet   20 millimeters = __2__ centimeters

2 feet = __24__ inches   20 decimeters = __2__ meters

3 feet = __1__ yards   30 centimeters = __3__ decimeters

2 yards = __6__ feet   2 meters = __200__ centimeters

36 inches = __3__ feet   3 centimeters = __30__ millimeters

1 yard = __3__ feet   200 centimeters = __2__ meters

Measurements

2 yards is greater than 2 feet.	(true) /	false
20 centimeters = 2 decimeters	(true) /	false
6 feet = 3 yards	true /	(false)
36 inches is less than 2 feet	true /	(false)
2 decimeters = 20 centimeters	(true) /	false
40 millimeters = 3 centimeters	true /	(false)
4 meters = 400 centimeters	(true) /	false

Measurement Abbreviation Review

pink   aqua   yellow

green   blue   red

yellow   aqua   green

purple   pink   red

Length Review

Across

1. __metric__ system uses m, cm, mm

3. 1 ft = 12 __inches__

6. 1 __centimeter__ = 10 millimeters

Down

1. 1 __meter__ = 100 cm

2. 1 __decimeter__ = 10 cm

4. 1 __yard__ = 3 feet

5. 3 __feet__ = 1 yard

Unit 15 Week 31 Answer Key

Standard Weight

1 ounce

   1 ton

      1 pound

         1 pound

           1 ton

              1 ounce

Comparing Weight

1	3	2
3	2	1
3	1	2
1	2	3
2	3	1

Ounces and Pounds

8 oz = __.5__ lbs   1 1/2 lbs = __24__ oz

24 oz = __1.5__ lbs   3 lbs = __48__ oz

32 oz = __2__ lbs   1/2 lb = __8__ oz

16 oz = __1__ lbs   2 lbs = __32__ oz

Rounding Weights

__32__ lbs   __13__ lbs

__15__ lbs   __12__ lbs   __8__ lbs

__2__ lbs   __20__ lbs

__3__ lbs   __11__ lbs   __25__ lbs

The Metric System

__2__ kg   __3__ kg   __2.5__ kg

__3.5__ kg   __0.5__ kg   __5__ kg

Adding Kilograms

__3.5__ kg	__10__ kg	__9__ kg
= 3,500 grams	= 10,000 grams	= 9,000 grams
__18__ kg	__10.5__ kg	__8.5__ kg
= 18,000 grams	= 10,500 grams	= 8,500 grams
__10__ kg	__13.5__ kg	__7__ kg
= 10,000 grams	= 13,500 grams	= 7,000 grams
__4.5__ kg	__7.5__ kg	__15__ kg
= 4,500 grams	= 7,500 grams	= 15,000 grams

Grams and Kilograms

90 kilograms

750 kilograms

3 kilograms

10 grams

160 kilograms

400,000 kilograms

Review of Weight Measurements

(pounds)   ounces   tons

kilograms   ounces   (tons)

pounds   tons   (grams)

tons   kilograms   (ounces)

grams   (tons)   ounces

Weight Review

<	>
>	<
>	<
>	<

Subtraction and Weights Review

__6.5__ pounds   __7.4__ grams

__29.4__ ounces   __1.3__ kilograms

__2.9__ tons   __1.8__ pounds

__12.5__ grams   __7.9__ ounces

Unit 15 Week 32 Answer Key

Measuring Capacity

2	1	3
3	1	2
1	2	3

Ounces, Cups, Pints, Quarts, and Gallons

__2__ gallons
__2__ pints
__12__ quarts
__2__ cups
__3__ quarts
__1__ pints
__8__ fluid ounces

Measuring Capacity

9 fl. oz.	4 fl. oz.	7 fl. oz.
1 fl. oz.	3 fl. oz.	2 fl. oz.
5 fl. oz.	0 fl. oz.	8 fl. oz.

Measuring Capacity

Measuring Capacity with Metric System

3490 ml → __3000 ml__ = __3 liters__	5650 ml → __6000 ml__ = __6 liters__
2843 ml → __3000 ml__ = __3 liters__	9262 ml → __9000 ml__ = __9 liters__
8370 ml → __8000 ml__ = __8 liters__	6485 ml → __6000 ml__ = __6 liters__
1199 ml → __1000 ml__ = __1 liter__	7732 ml → __8000 ml__ = __8 liters__
4910 ml → __5000 ml__ = __5 liters__	3553 ml → __4000 ml__ = __4 liters__

Milliliters and Liters

1 milliliter	1 cup	2 gallons	2 fluid ounces
4 pints	4 milliliters	6 quarts	6 gallons
1 quart	1 gallon	4 fluid ounces	1 gallon
1 liter	1 fluid ounce	8 quarts	1 milliliter
1 milliliter	12 fluid ounces	300 liters	3 pints

Adding Measurement

40 g	72 qt	12 fl. oz.
413 pt	73 c	592 L
543 ml	922 g	960 qt
979 L	760 pt	847 c

Subtracting Measurement

19 fl. oz.

150 gal

56 c

119 L

53 pt

Capacity Measurement Review

5400 milliliters = 5 liters and 400 ml	(True) False
1298 milliliters = 12 liters and 98 ml	True (False)
2050 milliliters = 20 liters and 50 ml	True (False)
3612 milliliters = 3 liters and 612 ml	(True) False
4481 milliliters = 4 liters and 481 ml	(True) False
7338 milliliters = 7 liters and 338 ml	(True) False
6529 milliliters = 65 liters and 29 ml	True (False)

Capacity Review

8 cups
2000 milliliters
1 quart
1 gallon
24 fluid ounces
12 pints
4 liters

3 cups
2 pints
2 quarts
6 quarts
16 cups
4000 milliliters
2 liters

Unit 16 Week 33 Answer Key

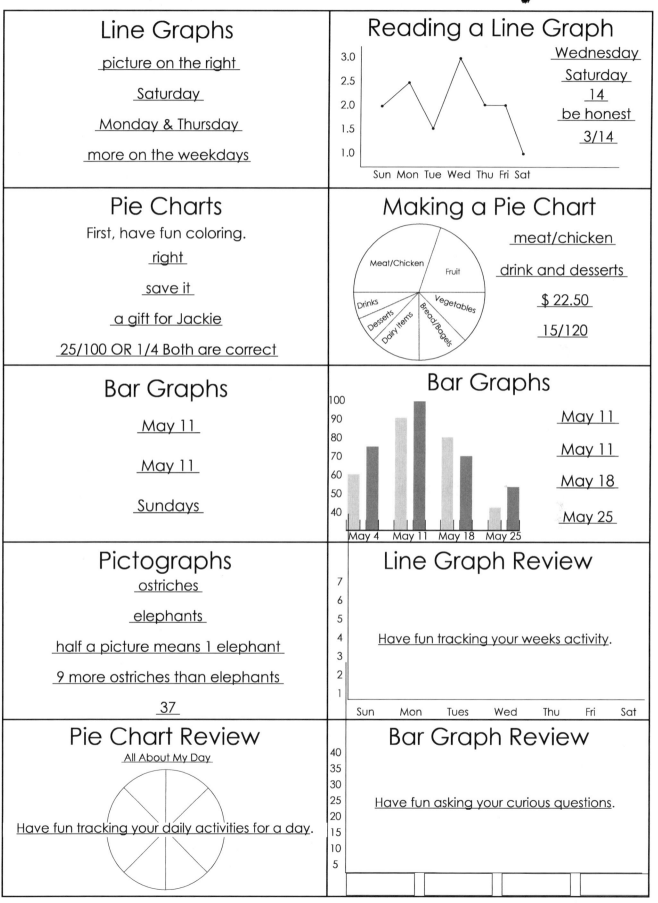

Line Graphs

picture on the right

Saturday

Monday & Thursday

more on the weekdays

Reading a Line Graph

Wednesday

Saturday

14

be honest

3/14

Pie Charts

First, have fun coloring.

right

save it

a gift for Jackie

25/100 OR 1/4 Both are correct

Making a Pie Chart

Meat/Chicken Fruit
Drinks
Desserts Vegetables
Dairy Items Bread/Bagels

meat/chicken

drink and desserts

$ 22.50

15/120

Bar Graphs

May 11

May 11

Sundays

Bar Graphs

May 11

May 11

May 18

May 25

Pictographs

ostriches

elephants

half a picture means 1 elephant

9 more ostriches than elephants

37

Line Graph Review

Have fun tracking your weeks activity.

Pie Chart Review

All About My Day

Have fun tracking your daily activities for a day.

Bar Graph Review

Have fun asking your curious questions.

Unit 16 Week 34 Answer Key

Parts of a Map

The United States

500

west

capital

Where is it Located?

The park is _south_ of Main Street. The bank is _north_ of the bridge.
      direction   direction

The gas station is on the corner of _Main Street_ and _Spring Street_.

Town hall is across the street from the ___church___.

The library is located on ___Spring Street___.

My house is on ___First Avenue___.

Traveling the World

Canada

Europe

Africa

Australia

What is your favorite location to visit?

Making a Map

Library	Lenny's house	grocery store
1st Street →		
fountain		fire station
2nd Street		
	$ Bank $	Pine Tree Park

Purple Place   Lemon Avenue

Purple   Yellow   blue

Reading a Grid

Green   Yellow   A ∏ ∐ A   Red

8 7 6 5 4 3 2 1
A B C D E F G H

Where are you?

School Supply Grid

A , 8   E , 6

F , 4   B , 3

A , 6   G , 8

C , 1   H , 1

Birthday Party Grid

B , 7   B , 1

D , 5   H , 2

E , 3   G , 6

Making a Grid

	A	B	C	D	E	F	G	H
8								chicken
7			sheep					
6		duck						
5					pig			
4						dog		
3								
2	horse							
1							barn	

Review of Maps and Grids

compass rose
    west
        letters
            numbers
                key
                    title
                        east
                            scale

Jesus and His Disciples Grid

Gray   Green   Jesus   Light Aqua   Pink
        Brown   Black   Yellow
                 Red   blue

8 7 6 5 4 3 2 1
A B C D E F G H I J K L

Unit 17 Week 35 Answer Key

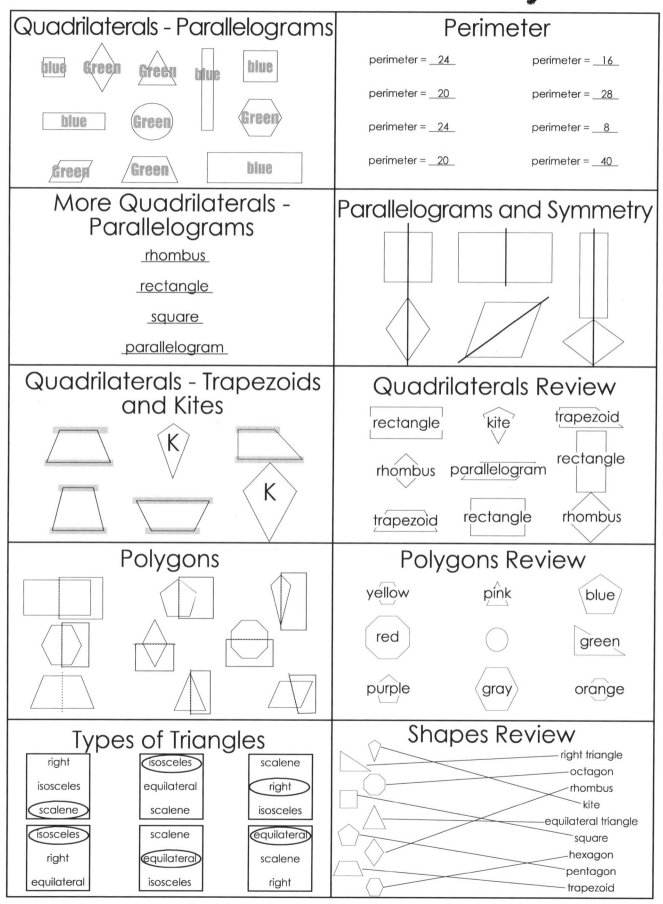

Quadrilaterals - Parallelograms

blue Green Green blue blue

blue Green blue Green

Green Green blue

Perimeter

perimeter = __24__   perimeter = __16__

perimeter = __20__   perimeter = __28__

perimeter = __24__   perimeter = __8__

perimeter = __20__   perimeter = __40__

More Quadrilaterals - Parallelograms

rhombus

rectangle

square

parallelogram

Parallelograms and Symmetry

Quadrilaterals - Trapezoids and Kites

K

K

Quadrilaterals Review

rectangle   kite   trapezoid

rhombus   parallelogram   rectangle

trapezoid   rectangle   rhombus

Polygons

Polygons Review

yellow   pink   blue

red   green

purple   gray   orange

Types of Triangles

right   isosceles   scalene
isosceles   equilateral   right
scalene   scalene   isosceles

isosceles   scalene   equilateral
right   equilateral   scalene
equilateral   isosceles   right

Shapes Review

right triangle
octagon
rhombus
kite
equilateral triangle
square
hexagon
pentagon
trapezoid

Unit 17 Week 36 Answer Key

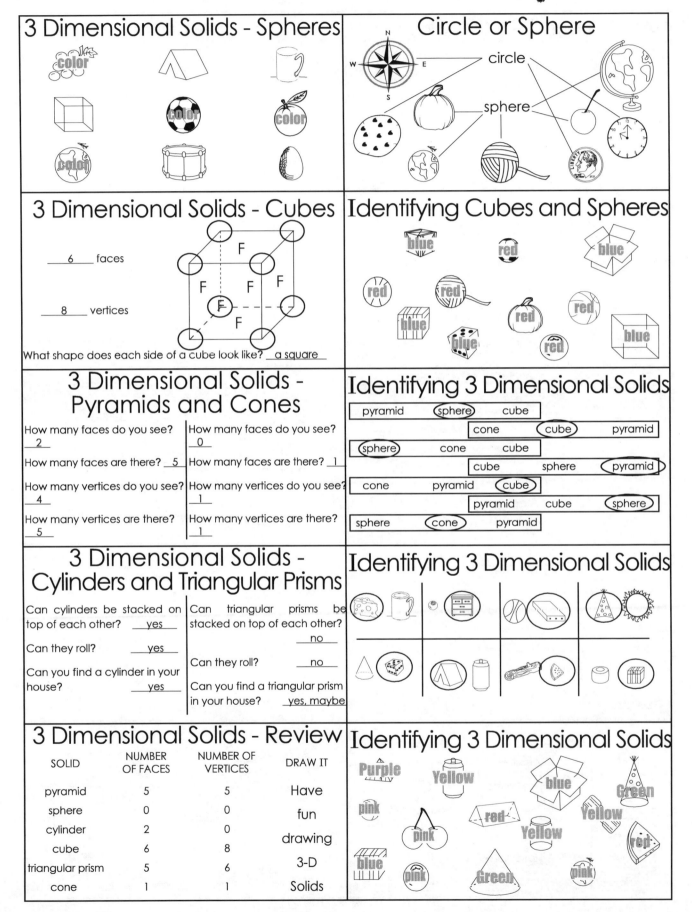

3 Dimensional Solids - Spheres

Circle or Sphere

circle

sphere

3 Dimensional Solids - Cubes

_6_ faces

_8_ vertices

F F F F F F

What shape does each side of a cube look like? _a square_

Identifying Cubes and Spheres

blue red blue
red red blue
blue red red blue

3 Dimensional Solids - Pyramids and Cones

How many faces do you see?
_2_

How many faces are there? _5_

How many vertices do you see?
_4_

How many vertices are there?
_5_

How many faces do you see?
_0_

How many faces are there? _1_

How many vertices do you see?
_1_

How many vertices are there?
_1_

Identifying 3 Dimensional Solids

pyramid	(sphere)	cube
cone	(cube)	pyramid
(sphere)	cone	cube
cube	sphere	(pyramid)
cone	pyramid	(cube)
pyramid	cube	(sphere)
sphere	(cone)	pyramid

3 Dimensional Solids - Cylinders and Triangular Prisms

Can cylinders be stacked on top of each other? _yes_

Can they roll? _yes_

Can you find a cylinder in your house? _yes_

Can triangular prisms be stacked on top of each other? _no_

Can they roll? _no_

Can you find a triangular prism in your house? _yes, maybe_

Identifying 3 Dimensional Solids

3 Dimensional Solids - Review

SOLID	NUMBER OF FACES	NUMBER OF VERTICES	DRAW IT
pyramid	5	5	Have
sphere	0	0	fun
cylinder	2	0	drawing
cube	6	8	3-D
triangular prism	5	6	Solids
cone	1	1	

Identifying 3 Dimensional Solids

Purple Yellow blue Green
pink red Yellow
pink Yellow red
blue pink Green pink

Resources

Place Value

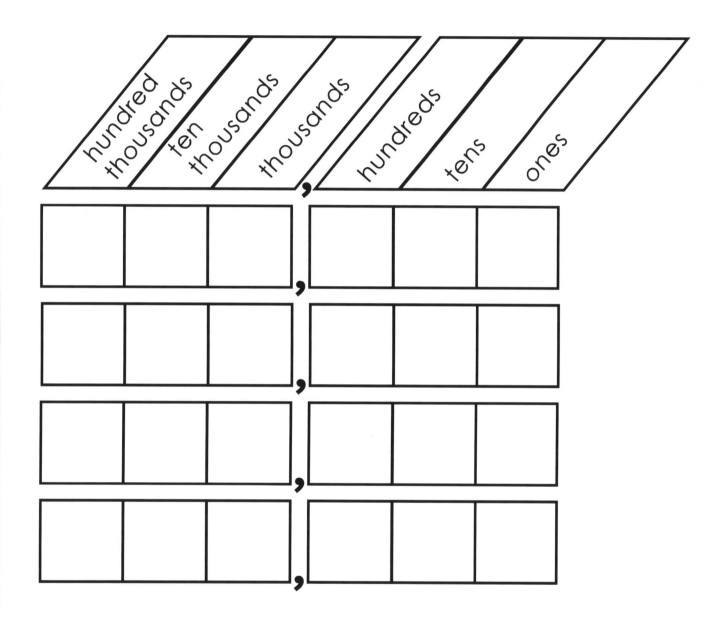

Remove from the book and laminate or place in a page protector. Practice writing in and reading numbers focusing on place value.

Analog Clock

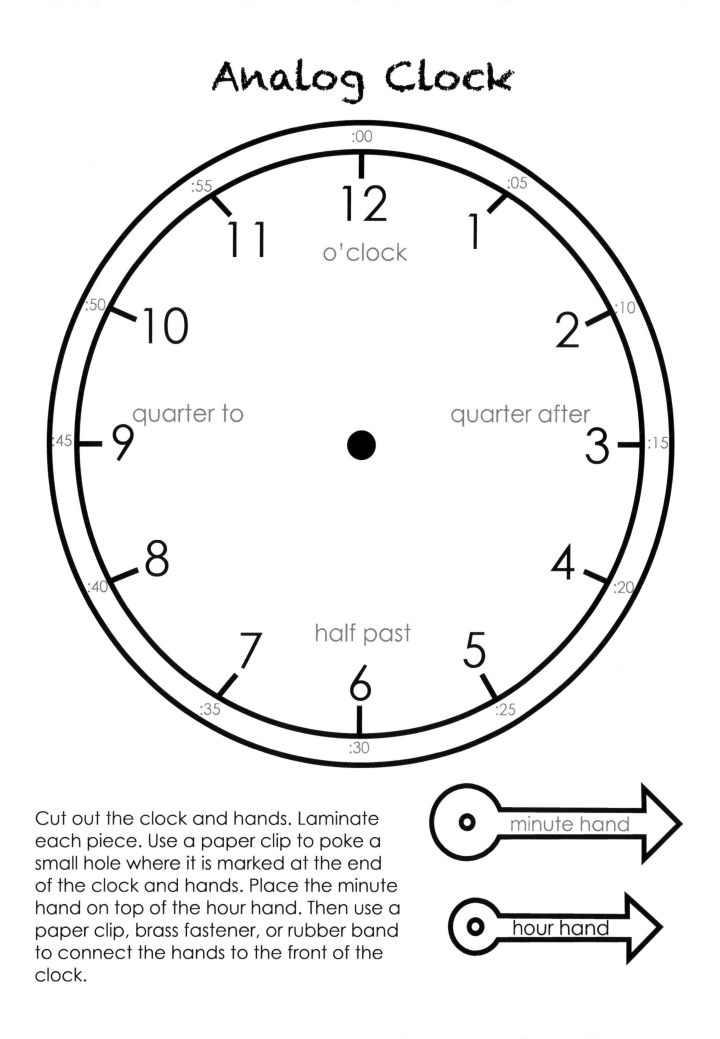

Cut out the clock and hands. Laminate each piece. Use a paper clip to poke a small hole where it is marked at the end of the clock and hands. Place the minute hand on top of the hour hand. Then use a paper clip, brass fastener, or rubber band to connect the hands to the front of the clock.

Thermometer

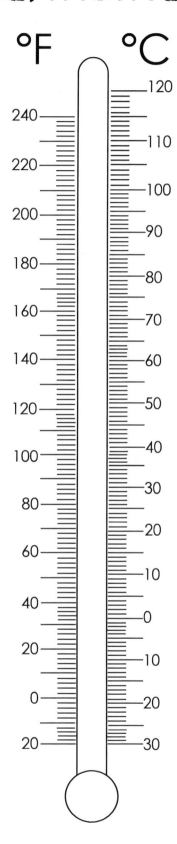

Remove from the book and laminate or place in a page protector. Practice marking and reading different temperatures.

Length Measurement Scales

Standard Length

12 inches = 1 foot   3 feet = 1 yard

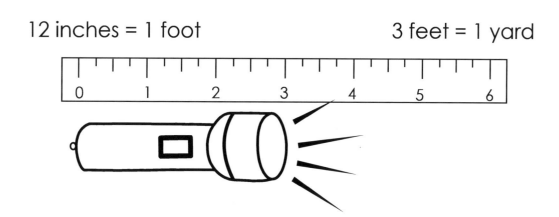

Metric Weight

10 millimeters = 1 centimeter   100 centimeters = 1 meter

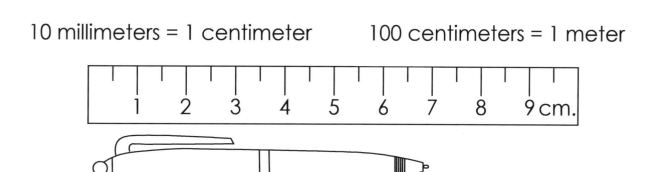

Weight Measurement Scales

Standard Weight

1 ounce 1 oz	16 ounces = 1 pound 1 lb	2000 pounds = 1 ton 1 t

Metric Weight

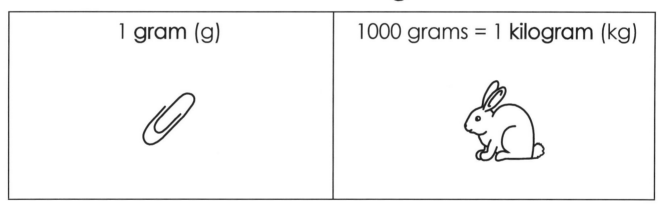

1 gram (g)	1000 grams = 1 kilogram (kg)

Capacity Measurement Scales

Standard Capacity

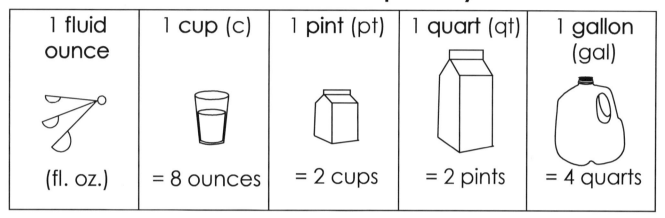

1 fluid ounce	1 cup (c)	1 pint (pt)	1 quart (qt)	1 gallon (gal)
(fl. oz.)	= 8 ounces	= 2 cups	= 2 pints	= 4 quarts

Metric Capacity

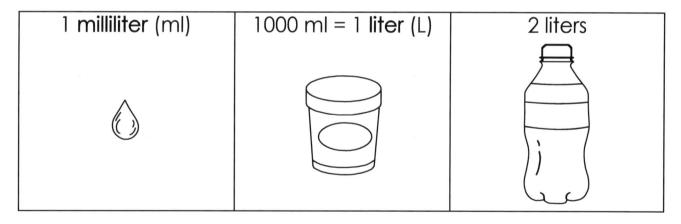

1 milliliter (ml)	1000 ml = 1 liter (L)	2 liters

2 - D Shapes

Quadrilaterals

1 set of parallel lines
trapezoid

2 sets of parallel lines: parallelograms

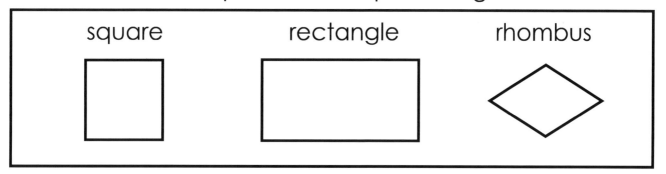

Polygons

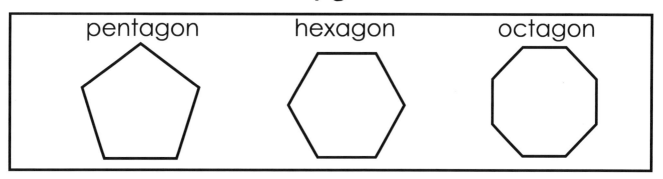

Triangles

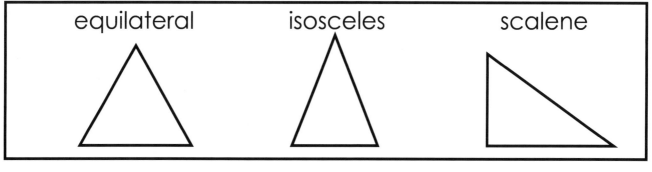

3 - D Solids

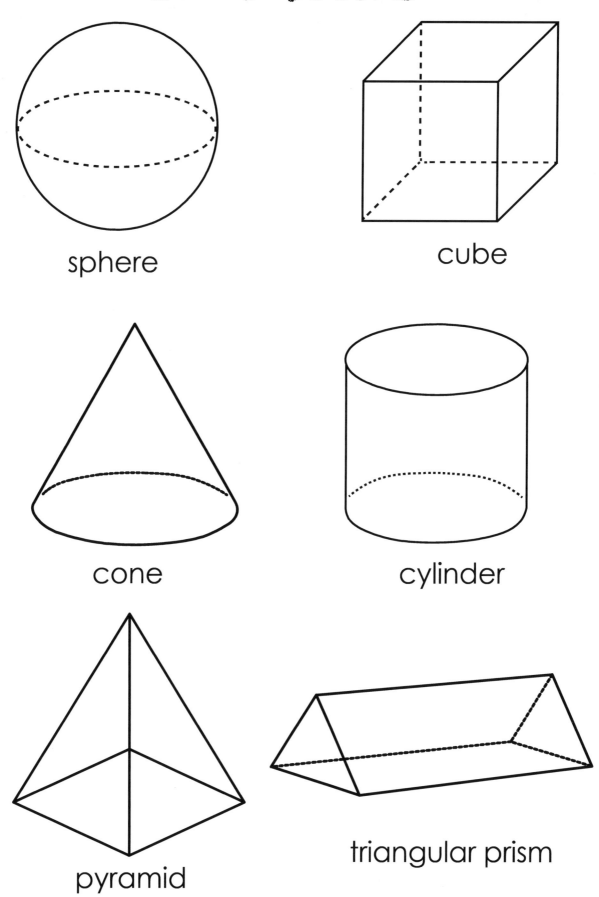

sphere

cube

cone

cylinder

pyramid

triangular prism